SOLID GEOMETRY

SOLID GEOMETRY

BY THE LATE

C. GODFREY, M.V.O., M.A.

AND

A. W. SIDDONS, M.A.

FORMERLY FELLOW OF JESUS COLLEGE, CAMBRIDGE;
AND LATE SENIOR MATHEMATICAL MASTER AT HARROW SCHOOL

CAMBRIDGE

AT THE UNIVERSITY PRESS

1949

CAMBRIDGE
UNIVERSITY PRESS

University Printing House, Cambridge CB2 8BS, United Kingdom

Cambridge University Press is part of the University of Cambridge.

It furthers the University's mission by disseminating knowledge in the pursuit of education, learning and research at the highest international levels of excellence.

www.cambridge.org
Information on this title: www.cambridge.org/9781316603857

© Cambridge University Press 1949

First edition 1911
Reprinted 1911, 1914, 1920, 1930, 1936, 1942, 1949
First paperback edition 2016

A catalogue record for this publication is available from the British Library

ISBN 978-1-316-60385-7 Paperback

PREFACE

"It is a great defect in most school courses of geometry
"that they are entirely confined to two dimensions. Even
"if solid geometry in the usual sense is not attempted, every
"occasion should be taken to liberate boys' minds from
"what becomes the tyranny of paper.......But beyond this
"it should be possible, if the earlier stages of the plane
"geometry work are rapidly and effectively dealt with as
"here suggested, to find time for a short course of solid
"geometry. Euclid's eleventh book is generally found dull
"and difficult, but all that is of real value in it can be dealt
"with much more rapidly, especially if full use is made of
"the idea of the motion of a line or of a plane. Similarly
"it should be found possible to include a study of the
"solid figures; this will be much facilitated if their general
"outlines have been made familiar at the very commence-
"ment as is usually the case."

> Board of Education Circular on the teaching of
> Geometry (No. 711, March 1909).

It may be argued that the course of plane geometry gives all
the practice necessary in the use of *formal* logic as applied to
mathematics, that the course in solid geometry should not aim
at giving further practice in formal logic, but rather at imparting
the power of 'thinking in space.' Whether this argument is
sound or not, it is generally found in actual practice that
the choice lies between informal solid geometry and no solid
geometry at all: there is no time for a course on the lines of
Euclid XI. This book is intended to provide an informal course:
a few theorems only are treated formally, mainly as illustrations of

the method to be used in solving such exercises as lend themselves
to Euclidean treatment.

The arrangement of the book is as follows:

Chaps. I.—VI.   An informal discussion of the main properties
of lines and planes.

Chaps. VII.—XIII.   Properties of the principal solid figures,
including mensuration.

> With an average class it may be better to take Chaps.
> VII.—XIII. before Chaps. I.—VI., referring back to the
> earlier chapters as occasion arises (e.g. when discussing the
> inclinations of faces of prism or pyramid, refer back to the
> section on the inclination of planes).

Chaps. XIV.—XVI.   Some account of coordinates in
3 dimensions, plan and elevation, perspective.

> The course of work in plan and elevation—sometimes
> called descriptive geometry—does not profess to give such
> technical skill as is needed by an architect or engineer.   On
> the other hand, it would be a good introduction to the
> subject for this class of student.   For the general mathe-
> matical student, this non-technical course is, we believe,
> both sufficient and necessary.   Under the influence of the
> new Tripos regulations at Cambridge, simple descriptive
> geometry must soon enter into the work of the higher
> classes at schools.   And, from an educational standpoint, it
> is perhaps the best possible subject for developing the space-
> imagination.

> If the aim is educational rather than technical, very
> accurate drawing is hardly necessary; whatever educational
> benefit is to be gained from drawing accurately has pre-
> sumably been gained during the earlier study of plane
> geometry.   But it must be remembered that in some cases
> a fair degree of accuracy is needed to reveal the essentials,
> e.g. there are often cases of concurrency or collinearity the
> failure of which would wreck the figure.

The teacher is recommended to illustrate his lessons with models made of paper, cardboard, string and needle-pointed sticks*; with the latter, especially, quite elaborate figures can be built up in a few seconds. The number of figures in the text has been kept down, on the assumption that 3-dimensioned models will be constructed.

The authors have to acknowledge with thanks the courtesy of Mr H. M. Taylor for permission to include some exercises from his edition of Euclid; of the Controller of H.M. Stationery Office for the use of exercises from Army papers, Science papers (Board of Education), and from the annual reports of the Secondary Education Department for Scotland; of the Director of Naval Education for the use of questions from various Navy papers.

Exercises that may be treated in a more or less formal and Euclidean style have been marked with a dagger (†).

* Such sticks as those supplied for the purpose by Mr G. Cussons, The Technical Works, Manchester.

C. G.
A. W. S.

October, 1910.

CONTENTS

† *indicates that an exercise is suitable for formal treatment, on Euclidean lines.*

CHAPTER I.

PLANES AND LINES.

The student will be familiar with the notion of **a plane**, of which we shall not offer a definition. The ordinary test for planeness is to apply a straight-edge to the surface in every direction, and see if it always fits. This test depends on the obvious fact that **if two points lie in a plane, the straight line joining them lies entirely in that plane.**

Ex. 1. How would the above fact be of use in testing a surface that is slightly convex?

Determination of a plane. It will be remembered that a straight line is fixed, or determined, by 2 suitable conditions. Thus it may be required to pass through two points, or to pass through a point and lie in a given direction.

We shall see that 3 suitable conditions are needed to determine a plane.

First, fix one point A of a plane (e.g. by placing the plane in contact with the corner of a table). The plane can now take up any direction, or **orientation**, in space. Fix a second point B: the plane can now turn about AB as a line of hinges. Fix a third point C, and the plane is deprived of all freedom of motion. The plane has thus been fixed by 3 conditions.

By experimenting with a sheet of cardboard, it will become clear that the above is only one of many ways of fixing a plane by suitable conditions, e.g.:

A plane is determined uniquely if it is required

(i)   to pass through three given points;

(ii)   to contain a given straight line and pass through a given point not on the line;

(iii) to contain two intersecting straight lines (or lines which intersect if produced)*;

(iv)   to contain two parallel straight lines (i.e. two straight lines intersecting at infinity).

Ex. 2. Try to make a plane contain two straight lines not intersecting nor parallel.

Ex. 3. Why does a three-legged table stand more steadily on uneven ground than a four-legged one?

Ex. 4. If each of three straight lines meets the other two, the three lines are either concurrent or coplanar.

Generation of a plane. A plane may be swept out, or 'generated'

(i) by a straight line passing through a fixed point and sliding on a fixed straight line, not containing the point;

(ii) by a straight line sliding on two fixed intersecting straight lines;

(iii) by a straight line sliding on two fixed parallel straight lines;

(iv) by a straight line sliding on a fixed straight line, and remaining in a fixed direction (or, moving parallel to itself). (This is really a particular case of (i), the fixed point being at infinity.)

Two planes intersect, in general, in a straight line; in other words, they determine a straight line.

Associate this with the fact that two points determine a straight line.

---

* In future it will be assumed that lines and planes are infinite, unless it is stated that a finite portion is under consideration.

If two planes have no finite point in common, they are parallel. Mathematically, their line of intersection is a line at infinity; a set of parallel planes determine one line at infinity.

Straight line and plane. In general, a straight line intersects a plane in a point (illustrate with a stick and the floor, etc.); that is to say, a straight line and a plane determine a point.

Compare this with the fact that a straight line and a point, in general, determine a plane.

If a straight line and a plane have no finite point in common, they are parallel. Mathematically, they have in common the point at infinity on the line.

Ex. 5. Hold a pencil parallel to a sloping desk, or book. Note that it need not be horizontal. Can it be horizontal? Can it be vertical?

Ex. 6. How many lines can there be through a given point parallel to a given plane? What can be asserted about the whole set of such lines?

Ex. 7. What is generated by a line passing through a given point, and moving so as to remain parallel to a given plane?

*Ex. 8. What is the locus of the intersection of a fixed plane, and a line through a fixed point intersecting a fixed line?

Ex. 9. What is the locus of the intersection of a fixed plane, and a line constant in direction and intersecting a fixed line?

Two straight lines. In general, two straight lines in space are not in the same plane, and do not intersect; such lines are said to be **skew** to one another.

If they happen to be in the same plane, they either intersect or are parallel.

Note carefully the distinction between skew and parallel lines. It has been pointed out in the course of plane geometry that parallel lines must be in the same plane. Parallel lines must also be in the same direction; skew lines cannot be so.

---

* In dealing with a mathematical problem, the student is strongly recommended not to pass on till he has examined the particular cases that may arise: e.g. in the present problem he should consider the case of the line being parallel to the plane.

4   SOLID GEOMETRY

Ex. 10. Find examples of (i) a pair of skew lines, (ii) a pair of parallel lines (not drawn on paper).

Ex. 11. How many lines are there in space through a given point and parallel to a given line?

Ex. 12. What can be asserted about the whole set of lines passing through a given point and intersecting a given line?

Ex. 13. Are *any* two horizontal lines necessarily parallel? Are *any* two vertical lines necessarily parallel?

Ex. 14. Find two horizontal lines skew to one another.

Ex. 15. Prove that, among the lines meeting two skew lines, it is impossible to find a pair which intersect or are parallel.

Ex. 16. Determine a line intersecting two skew lines and parallel to a third line.

Ex. 17. Given four straight lines of which two are parallel, one straight line can be drawn to intersect the four.

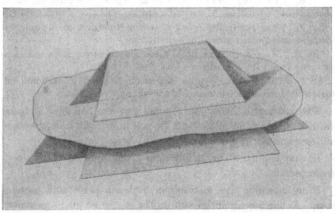

fig. 1.

Three planes. In general, any two planes have a line in common; this line intersects a third plane in a point. Three planes therefore determine a point (e.g. two meeting walls of a room, and the ceiling).

Compare these two facts—in general 3 planes determine 1 point ; 3 points determine 1 plane.

Particular case (i). It may happen that the line of inter-section of two of the planes is parallel to the third plane. The three planes will then have no point in common; see fig. 1. (Compare the three side-faces of a triangular prism; the three planes of a folding screen; the two slopes of a roof, and the ground plane.)

Notice that in this case the line common to *any* two of the planes is parallel to the third; the whole system will consist of three planes, and three parallel lines.

This case is derived from the general case by making the common point of the three planes retreat to infinity.

Particular case (ii). Two of the planes may be parallel. The third plane will cut these in two parallel lines, and there will be no point common to the three planes; see fig. 2. (Compare two parallel walls and the floor.)

This is a particular case of (i).

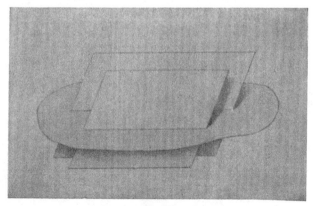

fig. 2.

Particular case (iii).   The common lines of each pair of planes may coincide; we shall then have three planes passing through the same line (like three leaves of a partly open book).   Here the three planes have an infinite number of common points, lying on a line.

Just as 3 points *may* lie in a line, so 3 planes *may* pass through a line.

Particular case (iv).   The three planes may be parallel (like the various floors of a building).

This is a particular case of (ii); also of (iii), for the three planes have in common a line at infinity.

Ex. 18.   How are the planes disposed that are drawn through a point to contain a set (i) of concurrent coplanar lines? (ii) of concurrent lines, not coplanar? (iii) of parallel coplanar lines? (iv) of parallel lines, not coplanar?

CHAPTER II.

PARALLEL POSITIONS OF PLANES AND LINES.

The reader should convince himself of the truth of the following statements, by placing books, pencils, etc. to represent planes and lines. The truth of a proposition is usually more evident if one of the planes in question is taken to be horizontal; most of the planes that we are concerned with in experience are horizontal or vertical.

(i) **Planes parallel to the same plane are parallel to one another.**

Ex. 19. What can be asserted about planes perpendicular to the same plane? about planes perpendicular to the same line? about planes parallel to the same line?

(ii) **Straight lines parallel to the same straight line are parallel to one another.**

Here note that the three lines need not be in the same plane, e.g. three edges of a triangular file. This property will be recognised as an extension of a property in plane geometry.

Ex. 20. What can be asserted about lines parallel to the same plane? about lines perpendicular to the same plane?

Ex. 21. Prove that the shadows of a number of vertical sticks, thrown on the ceiling by a candle, are a set of concurrent lines.

(iii) **A straight line parallel to a straight line lying in a certain plane is parallel to the plane.**

Ex. 22. How many straight lines can there be through a point parallel to each of two intersecting planes?

Ex. 23. In what case can there be one line parallel to each of three planes?

(iv) **If a plane α pass through a line l which is parallel to a plane β, the line of intersection of α and β will be parallel to l.**

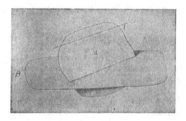

fig. 3.

Think of α as the sloping top of a desk, l the line of hinges parallel to β the ground-plane.

†**Ex. 24.** Prove this by *reductio ad absurdum*.

Ex. 25. In what case is a stick parallel to its shadow on the ground?

Ex. 26. If a plane α pass through a line l which is perpendicular to a plane β, what is the line of intersection of α and β?

Ex. 27. Let a line l meet a plane β obliquely, let α be a moving plane passing through l, and let m be the line of intersection of α and β. Prove that the different positions of m form a set of lines passing through a fixed point. Can any position of m be (i) parallel, (ii) perpendicular, to l?

Ex. 28. A line moves parallel to a given line so as always to intersect a given line; what does it generate?

Ex. 29. A set of planes are parallel to a given line. What is the arrangement of the intersections of these planes?

Ex. **30.** In a given plane, and through a given point in that plane, determine a line (i) to meet a given line (not in the plane), (ii) parallel to a given plane.

Ex. **31.** Through a given point determine a line parallel to a given plane and meeting a given line.

Ex. **32.** Through a moving point on a fixed line are constructed pairs of planes parallel to two fixed planes. What is the locus of the intersection of the planes so constructed?

(v) **Two parallel planes are cut by a third plane in parallel lines.**

(Geological strata generally lie in planes, over small areas. In such cases, their sections are seen as parallel lines on the sides of a railway cutting, the face of a cliff, etc.)

†Ex. **33.** Prove this by *reductio ad absurdum*.

Ex. **34.** What would be the shape of the section exposed by a slanting cut across a wooden plank?

(vi) **If intersecting lines l', m' are drawn respectively parallel to intersecting lines l, m, the plane containing l', m' is parallel to the plane containing l, m.**

(Draw l, m on a sheet of paper; and hold pencils to represent l', m'.)

(vii) **If intersecting lines l', m' are drawn respectively parallel to intersecting lines l, m, the angles between l', m' are equal to the angles between l, m.**

This will be recognised as an extension of a theorem in plane geometry.

(viii) **If straight lines are cut by parallel planes, they are cut in the same ratio.**

Data The straight lines AEB, CFD (fig. 4) are cut by the parallel planes α, β, γ (note that AEB, CFD are not necessarily in the same plane).

To prove that   AE : EB = CF : FD.

Proof   Let AD meet the plane β in X.

Join AC, BD, EX, XF.

The plane ABD cuts ∥ planes β, γ in ∥ lines EX, BD

∴ AE : EB = AX : XD.

Similarly, CF : FD = AX : XD.

∴ AE : EB = CF : FD.

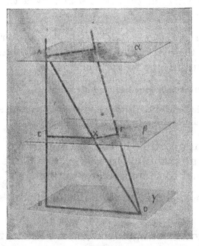

fig. 4.

Ex. **35.** In what case are E, X, F in a straight line? (fig. 4.)

Ex. **36.** In what case are AC and BD parallel? (fig. 4.)

†Ex. **37.** Points A, B, C, D in a plane are joined to a point O not in the plane. OA, OB, OC, OD are divided in the same ratio at A', B', C', D'. Prove that A'B'C'D' is a plane quadrilateral similar to ABCD. (Let plane A'B'C' cut OD in X ; shew that X coincides with D'.)

Ex. **38.** A stick 1 foot long is placed inside a box 10 inches high, touching the top and bottom. Prove that the locus of any point on the stick is a plane.

CHAPTER III.

PERPENDICULAR POSITIONS OF PLANES AND LINES.

It is obvious that if a line is perpendicular to a plane, it is perpendicular to any line that lies in that plane and meets it.

It is sufficient, however, if the line is perpendicular to *two lines* in the plane. This will be proved below; but the reader should first convince himself that it is not sufficient if the line is perpendicular to one line in the plane. Hold a pencil so as to meet a plane obliquely at a point, and see that there is bound to be one line in the plane and through the point which is perpendicular to the pencil.

Now take a piece of paper with a straight edge; fold this

fig. 5.

edge upon itself so as to form two right angles AOB, COB; open out the paper into two planes, and stand it upon the table with OA, OC on the table (see fig. 5). Then OB is perpendicular to OA, OC in the plane of the table; and is seen to be perpendicular to the table.

Ex. 39. Use two set squares to erect a line perpendicular to a plane.

(i) **If a straight line is perpendicular to two straight lines in a plane, at their point of intersection, it is perpendicular to any straight line lying in the plane and passing through the point of intersection.**

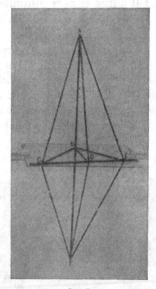

fig. 6.

Data   BA is ⊥ to BC, BE in plane *a*.
To shew that   BA is ⊥ to any straight line BD in the plane *a*.

Construction   Produce AB to F so that BF = AB.
Draw a straight line to cut BC, BD, BE in C, D, E.
Join C, D, E to A and F.

Proof In △s ABC, FBC, AB = FB, CB is common, ∠ ABC = ∠ FBC.
∴ AC = FC.
Similarly   AE = FE.
In △s ACE, FCE, AC = FC, AE = FE, CE is common.
∴ ∠ ACD = ∠ FCD.
In △s ACD, FCD, AC = FC, CD is common, ∠ ACD = ∠ FCD.
∴ AD = FD.
In △s ABD, FBD, AD = FD, AB = FB, BD is common.
∴ ∠ ABD = ∠ FBD.
∴ AB is ⊥ to BD.

(ii) **If a number of straight lines be drawn perpendicular to a straight line from a point in it, they all lie in a plane perpendicular to that line.**

This may be stated as a converse of (i) and proved by *reductio ad absurdum*; it may, however, be taken as self-evident.

Ex. **40.** If a right angle be rotated about one leg, what does the other leg generate?

†Ex. **41.** Shew that there cannot be a set of more than three lines, such that each is perpendicular to all the rest. (Assume that there may be four; and reduce *ad absurdum*.)

Ex. **42.** Explain how to use the 3, 4, 5 right-angled triangle to set a pole up vertically.

Ex. **43.** What can be asserted (i) about a set of straight lines perpendicular to a plane? (ii) about a set of planes perpendicular to a straight line?

Ex. **44.** In what case are a set of straight lines drawn perpendicular to a plane themselves coplanar (i.e. in one plane)?

Ex. **45.** How many planes are there through a given point (i) perpendicular, (ii) parallel, to a given line?

Ex. **46.** What is the condition under which there can be lines parallel to one given plane and perpendicular to another?

(iii) **Straight lines perpendicular to the same plane are parallel to one another.**

†Ex. 47. O is a point outside a plane *a*; AB is a line in the plane. In plane OAB, ON is drawn perpendicular to AB; in plane *a*, NP is drawn perpendicular to AB; in plane ONP, OP is drawn perpendicular to NP. Prove that OP is perpendicular to plane *a*.

†Ex. 48. OP is drawn perpendicular to a plane *a* from an external point O. From P, PN is drawn perpendicular to a line AB in *a*; shew that ON is perpendicular to AB.

Ex. 49. O is a fixed point outside a plane, and OA is a fixed line oblique to the plane. OP, perpendicular to OA, meets the plane in P. What is the locus of P?

The distance of a point from a plane is the length of the straight line drawn from the point perpendicular to the plane.

It is obvious that parallel planes are equidistant, and that the locus of points at a constant distance from a plane is a pair of planes parallel to the original plane.

Ex. 50. What is the locus of points in space at a constant distance from a straight line?

Ex. 51. What is the locus of points in space equidistant from two points? from two parallel lines? from two parallel planes? from two intersecting lines? from two intersecting planes?

Perpendicular planes. The meaning of the inclination of one plane to another will be explained later; in the meantime, it may be assumed that every one knows what is meant when one plane is said to be perpendicular to another.

Ex. 52. Shew how to set one plane at right angles to another by means of two set-squares.

Ex. 53. In what case does the section of two perpendicular planes by a third plane consist of two perpendicular lines?

Ex. 54. A plane revolves about a fixed line lying in itself (as the lid of a chest revolves about the line of hinges). Find a plane to which the revolving plane is always perpendicular.

Ex. 55.   What can be asserted about all the planes which pass through a given point and are perpendicular to a given plane?

Ex. 56.   A straight line moves so as always to meet a given straight line, and to be perpendicular to a given plane.   What does it generate?

Ex. 57.   What can be asserted about a set of planes perpendicular to each of two intersecting planes?

Ex. 58.   What condition must be fulfilled in order that there may be a plane perpendicular both to a given plane and to a given line?

Ex. 59.   How many planes can there be through a given point, parallel to a given line and perpendicular to a given plane?

Ex. 60.   Planes perpendicular to a given plane ; to what line are they all parallel?

†**Ex. 61.**   In a given plane through a given point in that plane draw a line such that the perpendiculars to it from two given points in space are concurrent.

†**Ex. 62.**   Through a given point determine a line such that the perpendiculars to it from three given points in space are concurrent.

CHAPTER IV.

OBLIQUE POSITIONS OF PLANES AND LINES.

Orthogonal projection*. The orthogonal projection of a point on a plane is the foot of the perpendicular from the point to the plane.

The orthogonal projection of a line (straight or curved) on a plane is the locus of the projections of the points of that line. It is clear that the projection of a straight line is, in general, a straight line; and that the projection of a finite straight line AB is the straight line joining the projections of A and B.

Ex. 63. In what case is the orthogonal projection of a straight line *not* a straight line?

Ex. 64. Shew that the orthogonal projection of a parallelogram is a parallelogram?

Ex. 65. In what case does a rectangle project into a rectangle?

Ex. 66. What is the relation between the base of a right prism and the figure disclosed by an oblique section of the prism?

---

* The term is merely explained here : for a treatment of the subject, see the authors' *Modern Geometry*.

Inclination of a straight line to a plane. The inclination of a straight line to a plane is defined to be the acute angle between the line and its orthogonal projection on the plane (fig. 7).

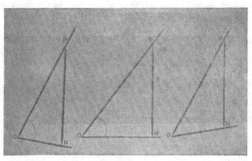

fig. 7.

Ex. 67. Prove that the length of the projection of a line on a plane = length of line × cos (inclination of line).

Ex. 68. A straight line AB meets a plane in A; and a straight line AC, lying always in the plane, revolves round A. Between what limits does the angle BAC vary?

†**Ex. 69.** A straight line CA always passes through a fixed point C, and moves so as to meet a fixed plane at a constant angle in A. Find the locus of A, and give a proof. What does CA generate?

†**Ex. 70.** Straight lines are drawn to meet two fixed parallel planes at a constant angle; prove that the part intercepted between the planes is of constant length.

†**Ex. 71.** A cylinder is cut obliquely by a plane. Prove that all the generators meet the plane at the same angle.

Ex. 72. Find the inclinations of a diagonal of a cube to the various faces.

†**Ex. 73.** Straight lines in a plane equally inclined to its common section with a second plane are equally inclined to the second plane.

Inclination of two planes. Take any point O on the line of intersection of the planes; draw OA, OB in the two planes at right angles to the line of intersection. The inclination of the planes (or the angle between these planes) is measured by the angle between these two lines. This angle is clearly independent of the position of O; if another pair of perpendiculars O′A′, O′B′ were drawn, these lines would be parallel respectively to OA, OB; and ∠ AOB = ∠ A′O′B′.

A cardboard model, such as that shewn in fig. 8, may be used to illustrate the inclination of two planes.

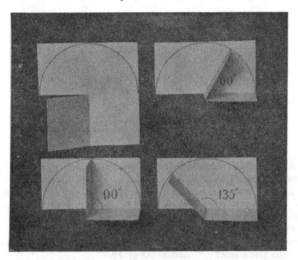

fig. 8.

Ex. 74. Consider the inclinations of the side-faces of an oblique prism. What plane figure has its angles equal to these inclinations?

Ex. 75. Open a book to a right angle: try if you can fit the acute angle of a set-square (or an acute angle formed by a pair of dividers) into the opening, so that the legs, though they include an acute angle, lie one on each of two planes at right angles. Given the position of the vertex of the acute angle, is the position of the set-square determinate?

Ex. 76. Mark on the surface of a tetrahedron lines giving the inclination of a pair of faces. Devise a method of ascertaining the size of this angle, without actually cutting the tetrahedron.

Ex. 77. Calculate the angle between two faces (i) of a regular tetrahedron, (ii) of a regular octahedron.

†**Ex. 78.** Prove that the angle between the planes of two meridians of the earth is equal to the angle between the tangents to the meridian circles at the Pole.

Ex. 79. In the cuboid of fig. 9, AE is drawn perpendicular to BC. Prove that plane DAE is perpendicular to BC. Hence find the inclination of plane BDC to the base of the cuboid.

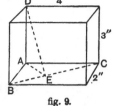

Ex. 80. A door, 8 ft. high and 3 ft. wide, swings through an angle of 40°. Calculate the angle between the new and old positions of a diagonal.

fig. 9.

Ex. 81. A 60° set-square is inclined at an angle of 40° to a horizontal plane; its longest side being in the plane. Find the inclination of each of the other sides to the horizontal plane.

Ex. 82. Fig. 10 represents a symmetrical roof; find the angle between the two planes that meet at the ridge.

Ex. 83. The height of the Pyramid of Cheops is 149 metres, and the base is a square of side 233 metres. Find the slope of a face, and the slope of an edge.

fig. 10.

Ex. 84. Two planes, inclined to the ground at an angle α, meet in an edge which slopes at an angle ϕ. Their ground-lines meet at an angle 2θ. Prove that $\tan \phi = \sin \theta \tan \alpha$.

Ex. 85. A rectangular looking-glass hangs on a wall in such a way that the bottom edge, 18″ long, lies against the wall, while the top edge is 3″ distant from the wall. The other edges are 12″ long. Find (i) the inclination of the glass to the vertical, (ii) the angle which a diagonal makes with the wall, (iii) the length of a line drawn through one of the lower corners to meet the upper edge, making an angle of 20° with a shorter side of the glass, (iv) the length of the projection of this line on the wall.

Ex. 86. A plane figure is drawn as follows :—OA and OB are straight lines at right angles to each other. On OA is drawn an isosceles triangle OAC, the angles at O and A being each 45°. On OB is drawn an isosceles triangle OBD, the angles at O and at B being each 60°. The sides OC and OD are each equal to 2 inches. The isosceles triangles thus formed are rotated about their bases OA and OB until OC and OD coincide. Find the angles which the planes OAC, OBD make with the plane OAB, and the distance from the plane OAB of the point where C and D coincide.

Ex. 87. ABCD is one square face of a cube, each edge of which is two inches in length ; E is the corner diametrically opposite A ; and F is the mid-point of CE. Draw on your paper a triangle equal to the triangle ABF.

Find by calculation the lengths of AF, BF, correct to two decimal places.

Find the tangent of the angle between the planes ABF, ABC.

2—2

CHAPTER V.

SKEW STRAIGHT LINES.

If **AB**, **CD** be two non-intersecting, or skew, straight lines, through any point on **AB** a line **PQ** can be drawn parallel to **CD**. All such parallels lie in a plane parallel to **CD**; we thus see that **through any straight line a plane can be drawn parallel to any skew straight line.**

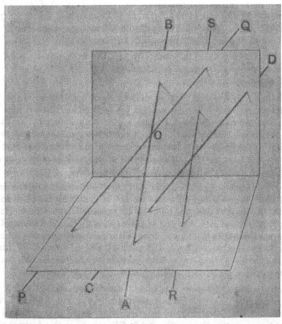

fig. 11.

This may be seen equally well by considering a plane through AB cutting CD in X, and then rotating this plane about AB till X goes to infinity.

If through any point on CD we draw RS parallel to AB, the plane determined by CD and SR is parallel to AB. And since AB, PQ are respectively parallel to RS, CD, the planes we have constructed through AB, CD are parallel to one another. Thus:

A pair of parallel planes can be constructed, each to contain one of two skew straight lines.

This principle suggests the following way of considering skew lines. Imagine the two lines and the two parallel planes to be turned round as a whole, till the planes are horizontal. The lines also will be horizontal, and in this position will be easier to deal with.

Ex. 88. Can (i) a vertical, (ii) a horizontal, plane be constructed to contain any given line?

Ex. 89. Shew that, through a given point, one line can be drawn to intersect two given skew lines. In what case is one of these intersections at infinity? Discuss also the particular case when the skew lines degenerate into intersecting lines.

Ex. 90. Shew that, in general, one line can be drawn parallel to a given line, to intersect two given lines. Discuss all special cases that arise.

Ex. 91. Determine a plane to cut two given planes in parallel lines, and *either* (i) to contain a given line *or* (ii) to pass through a given point and lie parallel to a given line.

†**Ex. 92.** Through a set of parallel lines are drawn a set of planes parallel to a fixed line. Prove that the set of planes meets a fixed plane in a set of parallel lines.

Angle between skew lines. The angles between two skew lines are defined to be the angles between one of the lines and a parallel to the second line intersecting the first line (e.g. ∠ BOP and ∠ BOQ in fig. 11).

Ex. 93. Why is this angle independent of the point through which the parallel is drawn?

Ex. 94. Give an instance of two skew lines at right angles; of three skew lines, each pair to be at right angles.

Ex. 95. What is the angle between skew diagonals of opposite faces of a cube?

Ex. 96. The edges of a cuboid are 2″, 3″, 4″. Find the angle between skew diagonals of the largest parallel faces.

Ex. 97. A line *a* turns about an axis *b*, skew to *a* and at right angles to it. What does it generate?

Common perpendicular to two skew lines AB, CD

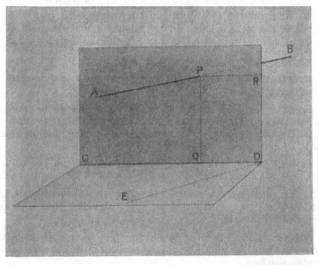

fig. 12.

To see that there must be a common perpendicular, imagine the two skew lines to be turned bodily into a horizontal position: no generality is lost by this process. They will then resemble a road crossed obliquely by a railway overhead (or, say, the centre

line of the road and the centre line of the railway). It is obvious
that one point of the railway is vertically above one point of the
road, that the two lines approach one another most closely at
these points, and that the vertical line joining these points is
a common perpendicular.

This suggests the following general construction :—

If AB, CD are the skew lines, through CD construct the plane
CDE ‖ to AB, and also the plane CDP ⊥ to CDE. Let plane CDP
cut AB in P ; and from P (in plane CDP) draw PQ ⊥ to CD.

Then PQ is the common perpendicular to AB and CD. For it
is ⊥ to CD by construction. To prove that it is ⊥ to AB, draw
PR ‖ to CD. Now PQ, lying in plane CDP and being ⊥ to CD, is
⊥ to plane CDE. And plane CDE is ‖ to plane BPR. Therefore
PQ is ⊥ to plane BPR. Therefore PQ is ⊥ to PB.

**The common perpendicular to two skew lines is the
shortest line connecting them.** For it is also perpendicular to
the two parallel planes containing the two lines, and is therefore
the shortest line connecting the planes. And any line connecting
the skew lines is also a line connecting the planes.

Ex. 98. What is the shortest distance between skew diagonals of
opposite faces of a cuboid? of a parallelepiped ?

†**Ex. 99.** Prove that the line joining the feet of the perpendiculars from
a point on two planes is at right angles to the intersection of the planes.

Ex. 100. Given the orthogonal projections of two skew lines on a
plane perpendicular to one of them, find that of their common perpendicular.

Ex. 101. Shew that the apparent crossing point of two skew lines
may be made to coincide with any two marked points on the lines, by
bringing the eye to a suitable position. What is the locus of the eye under
these conditions ?

Ex. 102. What is the condition that a plane may be constructed to
contain one line and to be at right angles to another skew line ?

A **skew quadrilateral** is the figure obtained by joining, in a certain order, four points not in a plane.

Ex. 103. If the vertices of a skew quadrilateral are joined in all possible ways, what figure is obtained?

Ex. 104. What is the locus of points viewed from which a skew quadrilateral looks like a plane angle?

Ex. 105. Explain how a skew quadrilateral may have four equal sides, two opposite angles right angles, and the other two angles acute.

Ex. 106. Explain how a skew quadrilateral may have only three of its angles right angles.

†**Ex. 107.** Shew that the figure obtained by joining the mid-points of adjacent sides of a skew quadrilateral is a parallelogram. Deduce a property of the two straight lines joining mid-points of opposite sides. Hence deduce a property of the tetrahedron.

†**Ex. 108.** Shew that if the opposite sides of a skew quadrilateral are equal, the opposite angles are equal.

Ex. 109. Shew how to draw a straight line through a given point, parallel to a given plane, and at right angles to a given line.

†**Ex. 110.** On two lines l, l' (in general skew to one another) are taken series of points P, Q, R, ... ; P', Q', R', ... such that

$$PQ : P'Q' = QR : Q'R' = \text{etc.}$$

Prove that PP', QQ', RR', ... are all parallel to a certain plane. Examine the particular cases in which (i) l, l' are parallel, (ii) l, l' are concurrent.

†**Ex. 111.** P, Q are variable points on two skew lines, and R divides PQ in a fixed ratio. Find the locus of R, and give a proof.

†**Ex. 112.** From a moving point P on a fixed line, PN is drawn perpendicular to a certain skew line. Shew that the velocities of P and N are proportional.

†**Ex. 113.** If P is a point on the line through the orthocentre of a triangle ABC and perpendicular to its plane, then PA is at right angles to BC.

CHAPTER VI.

LOCI.

The following propositions are self-evident:—

(1) **The locus of points at a given distance from a given point** is a spherical surface.

(2) **The locus of points at a given distance from a given plane** is a pair of parallel planes.

The reader will have no difficulty in proving the following:—

(3) **The locus of points equidistant from two given points** is the plane bisecting at right angles the line joining the points.

(4) **The locus of points equidistant from two intersecting planes** consists of the two planes containing the common section of the two given planes, and equally inclined to them.

It may be pointed out that, in general, the locus in space of points **determined by one condition** is a surface.

The locus of points **determined by two conditions** is, in general, the intersection of two surfaces, i.e. a line or lines, curved or straight.

The points **determined by three conditions** are the intersections of a line and a surface, i.e. are, in general, a finite number of points.

Ex. 114. What is the locus of points at a given distance from a given straight line?

Ex. 115. What is the locus of points equidistant (i) from two parallel lines? (ii) from two parallel planes? (iii) from two intersecting lines? (iv) from a plane and a line at right angles to it?

Ex. 116. What is the locus of points equidistant (i) from three points? (ii) from three planes meeting in a point? (iii) from three planes which are perpendicular to the same plane? (iv) from three lines intersecting in a point? (v) from three lines in a plane?

Ex. 117. What is the locus of a point P such that OP makes equal angles with fixed lines OA, OB?

Ex. 118. Investigate the properties of the tetrahedron that are analogous to the circumcircle, in-circle and ex-circle properties of the triangle.

Ex. 119. Find the locus of points at given distances from a given point and a given plane.

Ex. 120. What is the locus of the feet of perpendiculars from a fixed point upon planes (i) parallel to a fixed line? (ii) through a fixed line? (iii) through a fixed point?

†**Ex. 121.** A line through a fixed point O cuts two fixed planes in P, Q and moves so that OP : OQ is constant; R is taken on OPQ so that OP : OR is constant. Find the locus of R.

CHAPTER VII.

THE PRISM.

Ex. 122. From the figures below, or from examination of actual prisms, define

      (i) **a right prism,**

      (ii) **an oblique prism.**

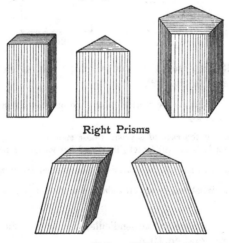

Right Prisms

Oblique Prisms
fig. 13.

Ex. 123. Describe how the side-faces of (i) a right prism, (ii) an oblique prism, may be generated by the motion of a finite straight line. Such a straight line is called **a generator.**

Ex. 124. Describe how (i) a right prism, (ii) an oblique prism, may be swept out by the motion of a plane area.

Ex. 125. What can be asserted as to the **ends** of a prism, namely the **base** and the **top**?

Prisms are commonly named according to the shape of the base, e.g. triangular, square, hexagonal, etc.

Ex. 126. What can be asserted as to the side-faces of (i) a right prism? (ii) an oblique prism?

Fig. 14 represents **a parallelepiped** (see also fig. 19).

Ex. 127. In what other terms may this solid be named?

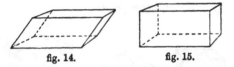

fig. 14.   fig. 15.

Fig. 15 represents a **rectangular parallelepiped** or **cuboid** (see also fig. 19).

Ex. 128. What can be asserted about the number and nature of the faces of a parallelepiped?

Ex. 129. In the case of a prism whose ends are *n*-gons, what is the number of (i) side-faces? (ii) faces? (iii) edges? (iv) corners or vertices?

Ex. 130. Of what classes of solids is a **cube** a particular case?

Ex. 131. What can be asserted about a **section** of a prism parallel to the ends?

A section of a prism perpendicular (normal) to the generators is called a **normal section**.

Ex. 132. The ends of a wooden prism are triangular and perpendicular to the sides, the lengths of the triangular edges being 13, 14, 15 cm. Calculate (i) the area of a normal section; (ii) the area of a section whose plane is parallel to the 15 cm. edge and at an angle of 30° to the axis.

Surface of prism.

Ex. 133. Prove that the surface* of a right prism = twice area of base + height × perimeter of base.

Ex. 134. Find the surface of the prism whose plan and elevation are given in fig. 16. (For 'plan and elevation', see Chap. xv.)

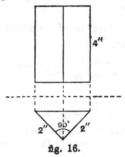

4″

2″ 90° 2″

fig. 16.

Volume of cuboid. (i) Let the length, breadth and height of a cuboid be a, b, c units of length, where a, b, c are integers.

The cuboid may be divided, by planes parallel to the base, into c slices, each one unit thick. The base of this slice may be divided into ab units of area; and the slice may be divided into the same number of units of volume, each unit of volume standing on a unit of area.

The cuboid has therefore been divided into $ab \times c$ units of volume; its volume is abc units of volume.

(ii) Let the edges of the cuboid be a, b, c units, and let a, b, c be not integers; e.g. let $a = 2\cdot4$, $b = 3\cdot55$, $c = 0\cdot362$. A sub-unit on the decimal scale can now be chosen, expressed in which the edges will have integral measure. Let the new unit be $\frac{1}{10^n}$ of the old; and let the new integral measures be a, β, γ; so that $a = 10^n . a$, $\beta = 10^n . b$, $\gamma = 10^n . c$. The volume is $a\beta\gamma$ cubic sub-units $= \frac{a\beta\gamma}{10^{3n}}$ cubic units $= abc$ cubic units.

---

* The whole surface is to be taken, including the two ends.

(iii)   The case of incommensurable edges is not discussed here, as presumably the student has not yet studied the meaning of multiplication of incommensurables and (ii) proves the rule to any degree of accuracy we please.

The volume of a cuboid is the product of the length, breadth and height.

Note also that **the volume of a cuboid is the product of the base-area and the height.**

Ex. 135. What is the ratio (i) of the surfaces, (ii) of the volumes of similar cuboids, the linear dimensions of the one being λ times those of the other?

Ex. 136. If the air-space in a room is twice that in a room of similar dimensions, compare the cost of papering the walls of the two rooms.

Ex. 137. By comparing two closed water-tanks, 6 ft. × 9 ft. × 4 ft. and 10 ft. × 8 ft. × 2 ft. respectively, shew that the tank which holds more water uses up, in this case, less sheet iron.

Ex. 138. Express (i) the edge, (ii) the diagonal, of a cube in terms of (a) the surface, (b) the volume.

Ex. 139. Express the surface of a cube in terms of the volume.

Volume of right prism.   Let the area of the base be S square units—say square millimetres; and the height h millimetres.

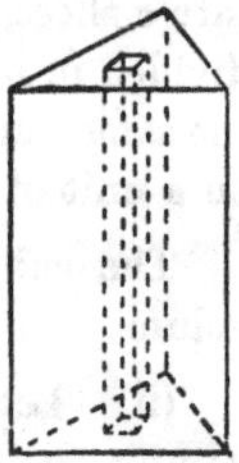

Suppose the base to be divided into square millimetres, e.g. by covering it with millimetre paper. On each little square stands a square pillar, h millimetres high.   The volume of this square pillar is h cubic millimetres; and as the base contains S square millimetres, the prism contains Sh cubic millimetres *.   Hence **the volume of a right prism is the product of the base-area and the height.**

fig. 17.

The prisms in fig. 18 have the same base-area and volume.

---

*   It will be noticed that this proof is really incomplete, as there will be broken squares at the edge of the base. This type of error can be made smaller without limit by taking smaller squares instead of square millimetres; and it can be shewn that the rule is strictly true. A strict proof involves the methods of the infinitesimal calculus.

The above reasoning applies to any right solid of uniform cross-section (e.g. rails for railways, T-irons, H-irons, cylinders, any metal goods made by rolling or pulling through a die).

The volume of a right solid of uniform cross-section is the product of the sectional-area and the height (or length).

fig. 18.

fig. 19.

Volume of oblique prism. A right prism can be built up with a number of thin cards, congruent with the base. If these cards are made to slide over one another as in fig. 19, an oblique prism will be formed, of the same base, height* and volume as the right prism. Hence the volume of an oblique prism is the product of the base-area and the height.

---

* The height of an oblique prism is the perpendicular distance between the ends.

†**Ex. 140.** Give an independent proof of the volume rule, in the case of a prism whose base is a right-angled triangle. Shew that any prism can be made up of prisms of this description, together with cuboids; hence give a general proof of the rule.

Ex. 141. Find the volume of the prism whose plan and elevation are given in fig. 16.

Ex. 142. A railway cutting 8 metres deep has to be made, with one side vertical and the other inclined at 30° to the vertical; the bottom is to be 9·4 metres broad. How many cubic metres (to the nearest integer) of material must be removed per kilometre?

Ex. 143. Sand lies against a wall, covering a strip of ground 4 ft. wide. If the sand will just rest with its surface inclined at 30° to the horizon, how much sand may lie on this strip per foot length of wall? Give the quantity to the nearest tenth of a cubic foot.

Ex. 144. A canvas tent has a floor a ft. square. Its top consists of a horizontal ridge b ft. high, the back and front are vertical, and the side walls are vertical for a height of c ft. from the ground. Find an expression for the area of the surface, and give a numerical result when $a=8$, $b=7$, and $c=2$. Also find the cubic content of a tent with these dimensions.

Ex. 145. Rain is falling steadily at the rate of 1 cm. per hour, and is caught by a trough formed of two equally inclined sides and two vertical ends. The depth of the trough is 20 cm. Draw a graph exhibiting the growing depth of the water, taking the time as abscissa in the scale of 2·5 cm. to the hour.

Shew that the graph is the same for all troughs of the same depth, whatever their lengths and angles, and find when the trough would be filled.

CHAPTER VIII.

THE CYLINDER.

Ex. 146. From fig. 20, or from examination of actual cylinders, define

      (i)   **a right cylinder,**

     (ii)   **an oblique cylinder.**

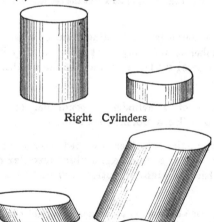

Right   Cylinders

Oblique Cylinders
fig. 20.

NOTE. A cylinder is sometimes conceived as **infinite**, i.e. having no plane ends but prolonged infinitely in each direction.

Ex. 147. Describe how the curved surface of

(i)   a right cylinder,

(ii) an oblique cylinder,

may be generated by the motion of a finite straight line. Such a straight line is called a **generator**.

Ex. 148. Describe how (i) a right cylinder, (ii) an oblique cylinder, may be swept out by the motion of a plane area.

Ex. 149. What can be asserted as to the **ends** of a cylinder, namely the **base** and the **top**?

Ex. 150. If the curved surface of a right cylinder is slit down a generator and opened out flat, what figure is produced?

Ex. 151. Make a rough sketch of the figure you would have obtained in the last example if the cylinder had been oblique, with circular ends.

Ex. 152. What can be asserted about the sections of a cylinder parallel to the ends?

Just as a curve is the limiting form of a polygon with an infinite number of infinitely short sides, so a cylinder is the limiting form of a prism with an infinite number of infinitely narrow side-faces.

A section of a cylinder perpendicular (normal) to the generators is called a normal section.

A right cylinder is further described according to the form of its base (or normal section), e.g. a right circular cylinder has a circular base; a right elliptical cylinder has an elliptical base, etc.

Ex. 153. Of what form is the section of a right circular cylinder by a plane parallel to the axis? Trace the change in this section as the plane moves parallel to itself.

Ex. 154. What is the shape of the shadow cast on the ground by (i) a circular sun-shade held horizontally, (ii) a sphere, the sun not being overhead?

The surface of a cylinder is usually taken to include the two ends. If the ends are excluded, the term **curved surface** is generally used.

Ex. 155. Prove that the surface of a right circular cylinder $= 2\pi r\,(r+h)$, where r is the radius of the base, h the height.

Ex. 156. Compare the side-surfaces of a right circular cylinder, and a square cuboid of the same height and the same base-area.

Ex. 157. The breadth of a garden roller is h ft., and the radius r ft. How many turns does it make in rolling an acre?

Ex. 158. Different cylinders are generated by the rotation of a rectangle about a side, according as it rotates about the long or the short side. Prove that the curved surfaces are the same.

Volume of cylinder. By the method of proof used for the prism, it may be shewn that the volume of a cylinder, **right or oblique, is the product of the base-area and the height.** The same result may be arrived at by regarding the cylinder as a limiting form of the prism.

Ex. 159. Prove that the volume of a right circular cylinder is $\pi r^2 h$.

Ex. 160. Express the surface of a right circular cylinder in terms of V, the volume, and h, the height.

Ex. 161. Compare the volumes of a right circular cylinder and a right square prism of equal height and equal girth.

Ex. 162. What percentage of wood must be lost in turning a rod of square section down to a cylinder?

Ex. 163. A rectangular sheet of zinc can be rolled into the curved surface of a cylinder in two ways, without overlap; which way gives the greater volume?

Ex. 164. A wrought iron plate $\frac{3}{8}''$ thick, measuring $3'$ by $1\frac{1}{2}$, has 20 rivet holes, each $1''$ in diameter. Find the weight of the plate (sp. gr. of wrought iron $= 7\cdot8$; 1 cu. ft. of water weighs $62\cdot4$ lbs.).

36   SOLID GEOMETRY

Ex. 165. The shaded rectangle in fig. 21 is
rotated about AB so as to generate a ring. Prove
that this ring has the same volume as a right prism
whose cross-section is the rectangle, and whose height
is the length of path described by the centroid of the
rectangle during a complete turn. Also prove that
the surface of the ring is the same as the side surface
of the prism. (This is a particular case of an im-

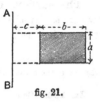

fig. 21.

portant theorem due to Pappus, who taught at Alexandria towards the end of
the 3rd century A.D.; the theorems were republished by Guldinus (1577–1643).)

Ex. 166. "To afford some notion of the present magnitude of the
petroleum trade, it may be stated that a pipe 41 inches in diameter would
be needed for the conveyance of the petroleum which the world is at present
using, assuming a rate of flow of 3 feet a second; and that for the storage
of a year's supply a tank 929 ft. in length, breadth and height would be
required." (*Encycl. Brit.* art. Petroleum, New Volumes, 1902.)

Check the consistency of the above statements by deducing a value of *π*.

Ex. 167. A hollow tube, open at both ends, 18″ long, 3″ external
diameter, is made of metal ⅛″ thick. It is closed by two equal cylindrical
caps, also of metal ⅛″ thick, which slip on at the two ends, like the lid of a
tin. The ends of the caps are flat.

Find the volume of either cap in terms of its length, *x″* (outside measure).
Find *x* if the weight of the two caps is half that of the tube.

Find also the total weight of tube and caps in the latter case, if the metal
used is brass weighing ·3 lb. per cubic inch.

CHAPTER IX.

THE PYRAMID.

Ex. 168. From fig. 22, or from examination of actual models, define a **pyramid**.

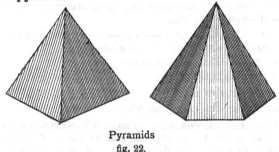

Pyramids
fig. 22.

Ex. 169. What figures are the **side-faces** of a pyramid?

A pyramid is called **regular** if its base is a regular polygon, and the line joining its **vertex** to the centre of the regular base is perpendicular to the base.

Pyramids are commonly named according to the shape of the base, e.g. triangular, square, hexagonal, etc.

Ex. 170. What are the side-faces of a regular pyramid?

†**Ex. 171.** Prove that a pyramid whose side-faces are isosceles triangles is not necessarily regular; but the base is a cyclic polygon.

Ex. 172. If the base of a pyramid is an n-gon, what is the **number** of (i) side-faces, (ii) faces, (iii) edges, (iv) corners?

Ex. 173. Find the height of a regular square pyramid whose base edge is 2″ and whose slant edge is 5″. Also find the slope of (i) a side-face, (ii) a slant edge

Ex. 174. The base of a regular pyramid is a regular hexagon of side *a*, and the slant edge is *b*; what is the height? Find the slope of (i) a side-face, (ii) a slant edge.

A pyramid whose base is a triangle (a triangular pyramid) is also called a **tetrahedron**, as it has four faces. A tetrahedron whose four faces are equilateral triangles is called a **regular tetrahedron**: the four faces are obviously congruent.

Ex. 175. Find the ratio of the height of a regular tetrahedron to its edge. Also find the angle between any two faces.

†**Ex. 176.** The perpendicular from a vertex of a regular tetrahedron upon the opposite face is three times the perpendicular drawn from its foot upon one of the remaining faces.

Ex. 177. A regular pyramid, vertex V, stands on a triangular base ABC, side 6″. Its height is 10″. Find the angle between

    (i)   The face VAB and the base ABC,

    (ii) The faces VAB and VBC,

    (iii) The edge VA and the base ABC,

    (iv) The edge VA and the face VBC.

Ex. 178. A regular pyramid, whose vertex is V, stands on a square base ABCD. Each side of base = 12 ft., and edge VA = 15 ft. P is a point in VA such that PA = 10 ft. O is the middle point of the square base ABCD. Find (i) distance OP, (ii) angle which OP makes with the face PAB.

If a pyramid is divided by a plane parallel to the base, the two pieces are respectively a pyramid and a **frustum of a pyramid**.

A section of a pyramid by a plane parallel to the base is similar to the base.

Proof It will be sufficiently general if we consider a pyramid OABCD, on a quadrilateral base.

    A plane parallel to the base cuts the pyramid in the section A′B′C′D′.

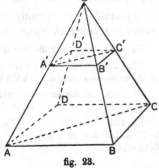

fig. 23.

The plane OAC cuts parallel planes A'B'C'D', ABCD in A'C', AC; which are therefore parallel lines. Similarly A'B', AB are parallel. Therefore ∠ B'A'C' = ∠ BAC. Similarly

$$\angle \text{B'C'A'} = \angle \text{BCA}.$$

△ A'B'C' is similar to △ ABC.

Similarly △ A'D'C' is similar to △ ADC.

Thus polygons A'B'C'D', ABCD may be cut up into corresponding and similar triangles.

The section is therefore similar to the base.

The pyramid OA'B'C'D' is clearly similar to the pyramid OABCD.

In accordance with the general principles of similarity :

(i) Any line associated with OA'B'C'D' is to the corresponding line associated with OABCD in the ratio of the linear dimensions of the pyramids (e.g. the heights of the pyramids are proportional to corresponding side edges, or to corresponding base edges).

(ii) Corresponding areas in the two pyramids are in the ratio of the squares of linear dimensions (e.g. the areas of A'B'C'D' and ABCD are proportional to the squares of the heights of the pyramids).

(iii) Corresponding volumes in the two pyramids are in the ratio of the cubes of linear dimensions (e.g. the volumes of the pyramids are proportional to the cubes of their heights).

The surface of a pyramid is usually taken to include the base.

Ex. 179. If the side-faces of a regular pyramid slope at any angle θ to the base, the total area of the side-faces is to the area of the base as sec θ : 1.

Ex. 180. Find the total surface of a regular pyramid of height h, whose base is a regular polygon of area S and inscribed radius r.

Volume of pyramid*. Let the area of the base be s and the height h.

Consider a thin slice of the pyramid cut off by two planes parallel to the base, distant x and $x + \Delta x$ from the vertex O. The thickness of the slice is Δx.

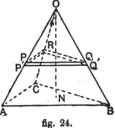

fig. 24.

The upper section is a polygon† similar to the base, its linear dimension being to those of the base as $x : h$.

The area of this face of the slice, therefore, is $\left(\dfrac{x}{h}\right)^2$ s.

The volume of the slice is $\left(\dfrac{x}{h}\right)^2$ sΔx, to the first order of small quantities.

The volume of the pyramid is $\displaystyle\int_0^h \left(\dfrac{x}{h}\right)^2 s\, dx$

$$= \frac{s}{h^2}\int_0^h x^2 dx$$

$$= \frac{s}{h^2} \cdot \frac{1}{3} h^3$$

$$= \tfrac{1}{3}\, sh.$$

Now the volume of a prism of the same base and height is sh. Hence the volume of a pyramid is $\frac{1}{3}$ of the product of base and height, i.e. $\frac{1}{3}$ of the volume of a prism of the same base and height.

This same result may be obtained without using the calculus, by means of the following theorems :

---

* Readers who have not studied the calculus are referred to the proof on page 41.

† In fig. 24 a triangular pyramid is shewn: but the proof applies to any pyramid.

(i)   Pyramids of equal height and on bases of equal area have the same volume.

(ii)   A triangular prism may be divided into three pyramids of equal volume.

Pyramids of equal height and on bases of equal area have the same volume.

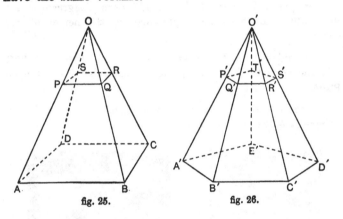

fig. 25.   fig. 26.

Let the height of the pyramids be h.

Let the pyramids stand on the same plane; and let a parallel plane distant x from either vertex cut the pyramids in PQRS, P'Q'R'S'T'. These polygons are similar to the bases; and their areas are to those of the bases in the ratio $x^2 : h^2$.

As the bases are equal in area, so are the corresponding sections.

If therefore the one pyramid be built up of a graduated set of thin cards, cards of equal area can be used to build up the other pyramid.

The two pyramids, therefore, have the same volume.

A triangular prism may be divided into three pyramids of equal volume.

The figure shews a triangular prism divided into three triangular pyramids. Any two of these pyramids may be shewn to have equal heights and equal bases, and therefore equal volumes.

For pyramids marked I and III have congruent bases PQR, ABC and the same height—namely the height of the prism.

Pyramids marked II and III have congruent bases PQA, BAQ and the same height—namely the perpendicular from C upon PQBA.

Hence the three pyramids are of equal volume, and each volume is $\frac{1}{3}$ that of the prism.

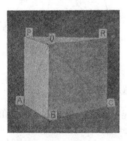 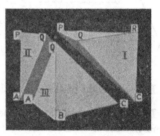

fig. 27.   fig. 28.

We have thus proved, in the case of pyramid QABC, that the volume of a triangular pyramid is $\frac{1}{3}$ of that of a prism on the same base, and of the same height.

This may be extended, at once, to any pyramid. Divide the base of the pyramid into triangles; then the pyramid can be divided into a number of triangular pyramids with a common vertex. If S_1, S_2, S_3, etc., be the areas of the triangles, the volumes of the triangular pyramids are $\frac{1}{3}S_1h$, $\frac{1}{3}S_2h$, etc., and the volume of the whole pyramid is $\frac{1}{3}h$ $(S_1 + S_2 + S_3 + ...)$, i.e. $\frac{1}{3}$ of the volume of the prism having the same base and height.

Ex. 181. In a rectangular block (fig. 29), M and N are the mid-points of two edges. Suppose that we remove the wedges with shaded ends MDC, MEF, leaving the wedge shown in fig. 30. S being the mid-point of MN, suppose two pyramids removed from fig. 30, so as to leave the pyramid shewn in fig. 31.

Taking the rectangular block to measure a by b by c as indicated in fig. 29, give an expression for the volume of the two wedges removed. Assuming that a pyramid has a volume of $1/m$ of the prism of the same base and height, give expressions for the volume of the two pyramids removed from fig. 30.

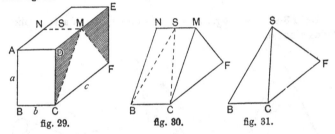

fig. 29.   fig. 30.   fig. 31.

Now get two expressions for the volume of fig. 31, one as $1/m$ of the original block, and the other as the difference between the original block and the pieces cut away. And find what value of m will make these expressions equal.

Ex. 182. A vessel containing water is in the form of an inverted hollow pyramid; its base is a square of side 6 ft., and altitude of pyramid is 10 ft. If the depth of the water is x ft., what is its volume?

Ex. 183. Cleopatra's Needle consists approximately of a frustum of a pyramid surmounted by a smaller pyramid. The lower base is $7\frac{1}{2}$ ft. square, and the upper base $4\frac{1}{2}$ ft. square; the height of the frustum is 61 ft. and of the upper pyramid $7\frac{1}{2}$ ft. Calculate the weight to the nearest ton, it being estimated that 1 cubic foot weighs about 170 lbs.

CHAPTER X

THE CONE.

Ex. 184. From fig. 32, or from examination of actual solids, define a cone.

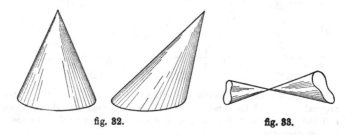

fig. 32.   fig. 33.

NOTE. A cone is sometimes conceived as **infinite,** i.e. having no plane base, but prolonged infinitely from the vertex. A still wider conception is that of the **infinite double cone,** see fig. 33.

Ex. 185. Describe how an infinite cone can be generated by the motion of a straight line. Such a straight line is called a **generator.**

Ex. 186. What can be asserted with regard to sections of a cone parallel to the base?

Cone as limiting form of pyramid. Just as a cylinder is the limiting form of a prism, so a cone is the limiting form of a pyramid.

Right circular cone. If the base of a cone is circular, and the **axis**, i.e. the line joining the vertex to the centre of the base, is perpendicular to the base, the cone is said to be a **right circular cone.** The generators in this case are all equal, and their length is called the **slant height** of the cone.

fig. 34.

Ex. 187. If the curved surface of a right circular cone is slit down a generator, and opened out flat, what figure is produced? How are the various measurements of this figure related to measurements of the cone?

Ex. 188. Describe how a right circular cone can be swept out by the rotation of a plane figure.

Ex. 189. A ring, diameter 24 ins., is suspended by six equal strings from a point 5 ins. above its centre, the strings being attached at equal intervals round its circumference. Find the angle between consecutive strings.

Ex. 190. What is the inclination of the generators of a right circular cone to the base, if the slant height is l, the diameter of the base d?

The section of a right circular cone by a plane through the axis is an isosceles triangle; the angle at the vertex of this triangle is called the **vertical angle** of the cone.

Ex. 191. Express the vertical angle of a cone in terms of two measurable dimensions of the cone.

Ex. 192. Describe in words the figures generated by the rotation of

(i) an obtuse-angled triangle about its longest side,

(ii) an obtuse-angled triangle about a shorter side,

(iii) a trapezium about the longer of the parallel sides,

(iv) a trapezium about the shorter of the parallel sides.

If a cone is divided by a plane parallel to the base, the two pieces are respectively a cone and a **frustum** of a **cone.** The conical piece is evidently **similar to the** whole cone.

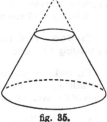

fig. 35.

Ex. 193. What is the section of a frustum of a cone by a plane containing the axis?

The surface of a right circular cone is usually taken to include the base. If the base is excluded, the term **curved surface** is generally used.

Ex. 194. By slitting the curved surface of a right circular cone along a generator, and developing it into a plane figure, shew that the area of this surface is half the product of the slant height and the circumference of the base.

Ex. 195. Find the whole surface of a cone of height h and base-radius r.

Ex. 196. Find, in terms of the curved surface of the original cone, the curved surface of a frustum of a cone whose height is $\frac{1}{n}$ that of the original cone. Use the principle of similarity.

Volume of cone. The volume of a pyramid is $\frac{1}{3}$ that of a prism of the same base and height. Proceeding to the limit, we see that the volume of a cone (of any shape) is $\frac{1}{3}$ that of a cylinder of the same base and height, and is therefore **one-third of the product of the base-area and the height.**

Ex. 197. Express the volume of a right circular cone in terms of the base-radius and the height.

Ex. 198. Express the volume of a right circular cone in terms of (i) height and semi-vertical angle, (ii) height and slant height.

Ex. 199. From the formulae obtained in Exs. 195, 197, verify the application to the cone of the rules mentioned on page 39, as to the ratios of surfaces and of volumes of similar solids.

Ex. 200. Find the surfaces and volumes of the three solids generated by rotating a triangle of sides 3", 4", 5" about its sides.

Ex. 201. A solid is generated by rotating a parallelogram of given base and height about its base. Shew that the volume is independent of the angle of the parallelogram.

Ex. 202. What area of canvas will be required for a conical tent, the diameter of the base being 20 feet and the vertical height 15 feet? How many cubic feet of air will the tent hold?

Ex. 203. A conical funnel rests inside a cylindrical jar as in fig. 36; the circumferences of the jar and the funnel-edge are 10″ and 14″ respectively, and the vertex of the funnel is 5″ below the rim of the jar. How much water would the funnel hold?

fig. 36.   fig. 37.

Ex. 204. A right circular cone is inscribed in a regular tetrahedron as in fig. 37; find the ratio of the volumes of cone and tetrahedron.

Ex. 205. A cone (height h, slant height l) is laid on its side on a plane, and rolled round so that the vertex remains fixed and the axis describes a second cone. Prove that the curved surface of this second cone is $\pi h^3/l$.

Ex. 206. A funnel made of sheet tin consists of a cylindrical part 4″ long attached to a frustum of a cone. If the total height is 8″, the radius of the cylindrical part $\frac{1}{4}$″, and the radius of the broad end $2\frac{1}{2}$″, find the area of tin required.

Ex. 207. Find the surface and volume of the frustum of a cone in terms of the radii of the ends (a and b) and the height h.

Ex. 208. A piece of paper in the shape of a large sector of a circle is rolled into a right circular cone; a series of cones may be so formed, according to the amount of overlap allowed in rolling. Find which of these cones has the greatest volume. What is the angle of the sector of paper? (Take r as radius of sector; xr as height of cone; form expression for volume, and find maximum graphically, or by differentiating.)

Ex. 209. Verify the theorem of Pappus (see Ex. 165) in the case of the solid generated by rotating an obtuse-angled triangle about its longest side.

CHAPTER XI.

THE SPHERE.

The reader should already know the meaning of the word **sphere.** Before attempting to define it, he should be warned that the word is commonly used in two different senses: sometimes as meaning a surface, sometimes the space enclosed by a surface. This does not, as a rule, cause inconvenience. Ambiguity may be avoided by speaking of **spherical surface** if necessary.

Ex. 210. Define a spherical surface.

Ex. 211. How may a sphere be generated by the rotation of a plane figure?

Ex. 212. What relation must hold between the radii of two spheres and the distance between their centres, in order that one may lie entirely within the other?

A sphere is symmetrical about any plane through the centre. Such a plane cuts the surface in a circle, called a **great circle,** whose radius is equal to that of the sphere.

fig. 38.

Ex. 213. Give instances of great circles on the surface of the earth.

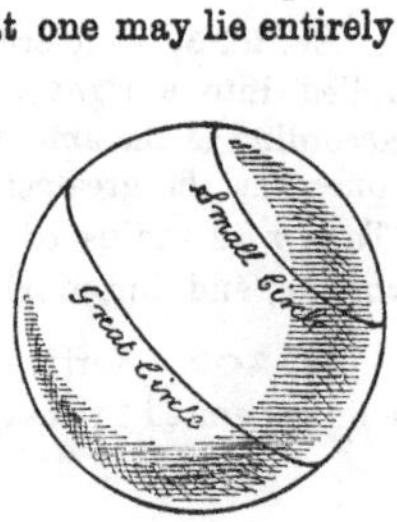

Plane and sphere. (i) A plane cuts a spherical surface, if the distance of the plane from the centre is less than the radius. The section is a circle; a great circle if the plane passes through the centre of the sphere; otherwise, a small circle, whose radius is less than that of the sphere.

(ii) A plane touches a spherical surface if the distance of the plane from the centre is equal to the radius; it is said to be a tangent plane. Touching may be regarded as a limiting case of cutting, the small circle having shrunk to a point-circle. Obviously the perpendicular from the centre to a tangent plane meets the plane at the point of contact, and at radius distance.

(iii) A plane does not meet a spherical surface if the distance of the plane from the centre is greater than the radius.

Ex. 214. Give instances of small circles on the surface of the earth.

†Ex. 215. Prove the above statement that if a plane cuts a sphere, the section is a circle.

Ex. 216. Find the area of a small circle whose plane is distant h from the centre of a sphere of radius r.

Straight line and sphere.

Ex. 217. Discuss the various positions that a straight line may have with respect to a sphere.

If a straight line touches a sphere, it is called a tangent line.

Ex. 218. Find the length of a chord of a sphere distant d from the centre.

Ex. 219. How many lines can be drawn to touch a sphere at a given point?

Ex. 220. TA, TB are tangents to a circle; the figure is rotated about the diameter through T; what is generated by (i) TA, TB, (ii) A, B, (iii) AB?

Ex. 221. The tangents to a sphere (radius r) from a point T, distant l from the centre, form a right circular cone : it is required to find the curved surface of this cone, bounded by its circle of contact.

Ex. 222. What is the locus of the mid-point of a chord of length $2l$, placed inside a sphere of radius r?

Ex. 223. The four corners of a square of side a lie on the surface of a sphere of radius r. What is the locus of the centre of the square?

Ex. 224. Find the radius of the horizon visible from the top of Snowdon (3560 ft.). [Take radius of Earth 4000 miles. Begin by finding the distance of horizon from point at height h, where h is small compared with R : prove that it is approximately $\sqrt{2hR}$.]

†**Ex. 225.** A sphere viewed from a point at a finite distance appears to be bounded by the circumference of a circle. Prove that this circle is a *small* circle of the sphere.

Ex. 226. Three balls, 5 cm. in diameter, lie on a floor in contact, and a fourth equal ball is placed on them. Calculate the height of the centre of the fourth ball above the floor.

†**Ex. 227.** Through a fixed point T is drawn a variable line, meeting the surface of a fixed sphere in A, B. Prove that TA.TB is constant; and discuss any particular cases that may arise.

Ex. 228. The altitude of the sun, seen from a ship at sea, is found to be a; and, knowing the Greenwich time, the navigator can determine the point on the earth which has the sun vertically overhead. What does he now know as to the position of the ship?

Two spheres that intersect may be shewn to intersect in a circle.

†**Ex. 229.** Prove the above statement by shewing that the curve of intersection lies in a plane. (Draw a perpendicular from a common point to the line of centres.)

Great circles. As the earth is approximately a sphere, the geometry of lines on a sphere is of practical importance, especially in navigation.

In general, **two points A, B on a sphere determine a great circle uniquely.** For a plane is determined by A, B, and the centre of the sphere; and this plane cuts the surface in a great circle. The points A and B can be joined by two great-circle arcs, the minor arc and the major arc of the great circle.

The **shortest line** on the surface of a sphere connecting two given points is the minor arc of the great circle through the points (this statement will be proved later); just as the shortest line in a plane connecting two points is the straight line.

Accordingly, in the geometry of the spherical surface—**spherical geometry** as it is called—the great circle takes the place of the straight line in plane geometry.

The **spherical distance** between two points on a sphere is the minor arc of the great circle through the points. It is generally measured in degrees of arc; i.e. the number of degrees which the minor arc subtends at the centre of the sphere.

The shortest-distance property of the great circle is used by navigators in **great-circle sailing**. Melbourne is approximately in the same latitude as Cape Town; but to sail along the parallel would be a much longer voyage than the great-circle route; this would run far to the South, as may be verified by stretching a piece of cotton on a globe between the two points. It is, in fact, impossible to follow the great-circle route throughout this voyage as it runs into the high latitudes, where navigation is unsafe.

Again, the great-circle route from the West coast of Mexico to Japan runs up the coast of America past San Francisco.

†**Ex. 230.** Prove that an infinite number of small circles can be drawn through two given points on a sphere.

†**Ex. 231.** Prove that any two great circles bisect each other at two diametrically opposite points.

Ex. 232. Find an exception to the rule that two points on a sphere determine one great circle.

Ex. 233. An arc of a great circle subtends an angle of θ radians at the centre of a sphere of radius r. What is its length? What is the length of an arc of x degrees?

If two points A, B on a sphere are **at opposite ends of a diameter**, it is no longer the case that only one great circle can be drawn through them. It is possible to draw an infinite number of planes containing the diameter, and each one of these cuts the surface in a great circle. We see therefore that through diametrically opposite points on a sphere an infinite number of great circles can be drawn (e.g. the meridians of the earth, drawn through N. and S. poles).

Poles of a great or small circle. The diameter of the sphere perpendicular to the plane of a great or small circle cuts the spherical surface in two points, called the **poles** of the circle; e.g. the North and South poles of the earth are the poles of the equator and of the parallels of latitude.

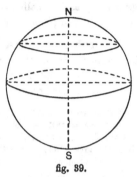

fig. 39.

The angles between two intersecting lines on a sphere are defined to be the angles between the tangents to these lines at their points of intersection.

†**Ex. 234. Prove that the angles between two great circles are equal to the angles between their planes.**

†**Ex. 235.** Prove that a great or small circle is at right angles to all the great circles through its poles.

†**Ex. 236.** Prove that the acute angle between two great circles is equal to the spherical distance between their poles.

†**Ex. 237.** Prove that the great circle through the poles of two great circles has for its poles the intersections of these circles.

Ex. 238. A small nail is driven into the highest point of a smooth spherical globe of radius r. A hoop, radius a, of fine wire, hangs on the nail and rests on the surface of the sphere. What angle does the plane of the hoop make with the vertical? What is the radius of the horizontal circle on the sphere which just touches the lowest point of the hoop?

Coordinates on a sphere. Latitude and longitude are simply coordinates on the surface of a sphere.

The equator and the meridian of Greenwich, NAS (fig. 40), are taken as rectangular great-circle axes; through the point P is drawn a great circle NPM, at right angles to the equator and therefore a meridian; AM and PM, measured in degrees, are the longitude and latitude. Longitude is measured E. and W. of A through 180°; latitude is measured N. and S.

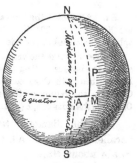

Ex. 239. What is the latitude of the Equator? of the Poles?

Ex. 240. Supposing it possible to travel along a great circle from a point 70° N. 90° E. to a point 70° N. 90° W., find the ratio of the length of the great-circle route to that of the route travelling due east.

Ex. 241. Find the distance along a parallel between points P and Q in the same

fig. 40.

latitude θ North, longitudes respectively α and β East. Also find the great-circle distance as follows: let N be the centre of the small circle, O of the sphere; compare the triangles PQN, PQO; hence prove that

$$\angle POQ = 2 \sin^{-1}\left(\sin\frac{\alpha-\beta}{2}\cos\theta\right).$$

Hence find numerically the ratio of the small-circle distance to the shortest distance when $\theta = 60°$, $\alpha - \beta = 40°$; also the distance saved by great-circle sailing, taking $r = 4000$ miles.

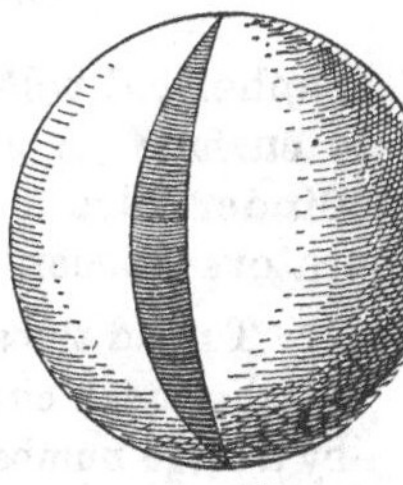

Lune. The portion of a spherical surface enclosed by two half great circles joining diametrically opposite points on the surface is called a **lune**.

Ex. 242. Find the ratio of the surface of a lune of angle α to the whole surface of the sphere.

Spherical triangle, polygon. If

fig. 41.

three or more points on the surface of a sphere are joined, in order, by minor arcs of great circles, the figure so formed on the surface is called a **spherical triangle** or **spherical polygon**. The sides of a spherical polygon are the spherical distances between the points, and are usually measured in degrees of arc; the **angles** of a spherical polygon are the angles between the great circles.

Ex. 243. Find the sides and angles of a spherical triangle whose vertices are 90° N.; 0° N., 30° W.; 0° N., 60° E. What fraction of the surface of the sphere is enclosed in the triangle?

Ex. 244. O is the vertex of a pyramid on a square base ABCD. Each side of the base is 8″ long, and the height of the pyramid is 12″. A sphere is described with O as centre, and A, B, C, D are points on the surface. Calculate the radius of the sphere. A, B, C are joined by great circles. Calculate the sides *a*, *b*, *c* (in degrees), and the angle B of the spherical triangle.

Let the diameter NS (fig. 42) be at right angles to the parallel chords AB, A'B'; and let the figure rotate about NS.

fig. 42.

The circle generates a sphere; A, B a small circle; AB, A'B' parallel planes; the arcs ANB, ASB **caps** of the spherical surface, or **spherical caps**; the arcs AA', BB' a **spherical belt**; the segments ANB, ASB **spherical segments**; the sectors AOBN, **AOBS spherical sectors.**

MENSURATION OF SPHERE.

A spherical belt cut off by two parallel planes has the same surface as a belt cut by the same planes from a cylinder circumscribing the sphere and having its generators perpendicular to the planes. (See fig. 43.)

Proof* To find the surface of the spherical belt bounded by the parallel small circles AB, A'B' (fig. 44), let the belt be divided by a large number of parallel planes into narrow belts such as that bounded by small circles PQ, P'Q'. Let N, S be the poles of the small circles; O the centre of the sphere. Each small circle subtends a cone at the centre of the sphere, whose semi-vertical angle will serve to determine the circle; e.g. small circle PQ is determined by $\angle NOP = \theta$; the neighbouring circle P'Q' by $\theta + d\theta$; AB, A'B' by a, a' respectively.

---

* For proof without calculus see p. 57.

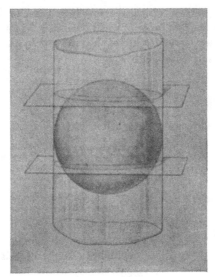

fig. 43.

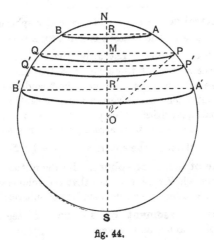

fig. 44.

The surface of the narrow belt bounded by PQ, P'Q' may, to the first order of small quantities, be found by multiplying the circumference of the circle PQ by the arc PP'.

∴ surface of narrow belt

$$= 2\pi \cdot \text{PM} \cdot \text{PP}'$$
$$= 2\pi r \sin \theta \cdot r \, d\theta$$
$$= 2\pi r^2 \sin \theta \, d\theta.$$

∴ surface of finite belt

$$= \int_a^{a'} 2\pi r^2 \sin \theta \, d\theta$$
$$= 2\pi r^2 \left[-\cos \theta \right]_a^{a'}$$
$$= 2\pi r^2 \left(\cos a - \cos a' \right)$$
$$= 2\pi r \left(\text{OR} - \text{OR}' \right)$$

= circumference of normal section of cylinder × height of belt of cylinder cut off

= surface of corresponding belt of cylinder

A spherical cap is a particular case of a spherical belt.

The surface of the sphere is $2\pi r \times \text{NS} = 4\pi r^2$.

Volume of sphere. Let σ be the surface of any small region on the surface of the sphere; S the whole surface. A radius whose extremity travels round the boundary of this region will sweep out a cone whose vertex is at the centre. To the first order of small quantities, we may regard this as a cone of height r standing on a plane base of area σ; its volume is $\frac{1}{3} r \sigma$.

Then the volume of the sphere $= \Sigma \cdot \frac{1}{3} r \sigma = \frac{1}{3} r \cdot \Sigma \sigma = \frac{1}{3} r S = \frac{4}{3} \pi r^3$.

Volume of sector of sphere. By the method of reasoning applied to the whole sphere we see that the volume of a sector of a sphere is $\frac{1}{3}$ × radius of sphere × surface of spherical cap.

Volume of segment of sphere. A segment is the difference of a sector and a cone

The **surface of a sphere** may be found **without the use of the calculus** by the following steps, the proof of which is left to the reader.

(1) The curved surface of a frustum of a cone is the product of the slant height and the mean of the circumference of the two ends. (This is another case of Pappus' theorem, see Ex. 165.)

(2) The surface generated by the rotation of a finite straight line about a coplanar axis is $2\pi \times$ projection of line on axis $\times$ perpendicular to line at mid-point terminated by axis.

(3) A circle may be regarded as the limit of a polygon. Rotating the circle about a diameter, we see that a sphere may be regarded as the limit of a figure composed of shallow truncated cones. A finite belt of a sphere may, in fact, be divided by a large number of parallel planes into narrow belts which may be treated as shallow truncated cones. Applying (2), we see that in the limit the surface of a finite belt of a sphere is $2\pi \times$ intercept on axis of sphere by planes bounding the belt $\times$ radius of sphere.

Hence the relation between surface of sphere and of circumscribing cylinder.

Ex. 245. Prove that half the earth's surface is included between parallels 30° N. and 30° S.

Ex. 246. Find the ratio of (i) the surfaces, (ii) the volumes, of a sphere and the circumscribed cube.

Ex. 247. A cube and a sphere have the same volume; find the ratio of their surfaces.

Ex. 248. To double the gas-capacity of a spherical balloon, in what ratio must the area of material be increased?

Ex. 249. A boiler has the form of a right circular cylinder with two convex hemispherical ends. Shew that the area of its external surface is equal to the product of its greatest length and the circumference of the circular section of the cylinder.

Ex. 250. Find the volume of a lens in the shape of a spherical segment, from the following measurements : thickness $= 0.22''$, diameter $= 1.65''$.

58 SOLID GEOMETRY

Ex. 251. Prove that the volume of a segment of a sphere of height h (i.e. cut off by a plane distant $r \sim h$ from the centre) is $\pi h^2 \left(r - \dfrac{h}{3} \right)$.

Ex. 252. Find an expression for the volume of metal in a hollow metal shell, spherical inside and out. If the thickness t is small compared with the external diameter $2r$, deduce that the volume of metal is approximately $4\pi r^2 t$.

Ex. 253. A piece of lead piping 8 ft. long, having an external diameter of 1 inch and an internal diameter of ·75 of an inch, is melted down and formed into 100 spherical shot of equal size. Find the diameter of one of the shot.

Ex. 254. In considering the volume of an oblique pyramid, we have assumed that two solids have the same volume if, being placed in the correct relative position, they are cut by any plane parallel to a certain plane in two sections of equal area. Apply this principle to find the volume of a hemisphere, by comparing it with a solid obtained by subtracting a cone from a cylinder. (See fig. 45.)

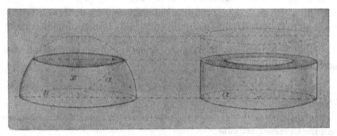

fig. 45.

Ex. 255. A golf ball is found to float immersed to the depth of $\frac{9}{16}$ of its diameter. What is its specific gravity?

Ex. 256. Segments are taken of two spheres, diameter 14 ins. and 7 ins. respectively, the heights being inversely as the diameters; if the volumes of the segments be equal, shew that the area of the spherical surface of each segment is very nearly 132 sq. ins.

Ex. 257. What proportion of the surface of a sphere (radius r) is visible from a point distant l from the centre?

Ex. 258. AB is a diameter of a given circle, and AC is a chord inclined 30° to AB. The area bounded by AC, AB and the arc BC revolves about AB. State the volume swept out by this area, and the area of the surface generated in terms of the radius of the circle.

Ex. 259. A cylindrical hole of diameter $2c$ is drilled through a solid sphere, diameter $2a$, the axis of the cylinder passing through the centre of the sphere. Shew that the volume of the remaining portion of the sphere is

$$\frac{4\pi}{3}(a^2 - c^2)^{\frac{3}{2}}.$$

Ex. 260. A spherical segment is 1 metre high, and the radius of its base is 2 metres. This segment is cut into two pieces by a plane parallel to its base and 50 cm. from it. Find the radius of the sphere and the area of the section.

Ex. 261. The internal diameter of the rim of a spherical basin is 6 inches, and the depth of the basin is 4 inches. Three equal balls float in the basin, which is full of water, the balls being just immersed. Determine the greatest diameter which the balls can have when they are placed symmetrically in the basin.

Draw a diagram shewing the basin and the balls in plan and elevation.

Ex. 262. A circle can be drawn to pass through any three points, not collinear. What is the corresponding theorem for a sphere?

†Ex. 263. Prove that a sphere may be described through any two circles which intersect in two points. Discuss the particular case of coplanar circles.

What are the conditions that a sphere can be described through two non-intersecting circles in space?

Ex. 264. The radius of a spherical surface (e.g. a lens) is measured by means of a spherometer, the theory of which is as follows: The spherometer has four legs, and, when it is resting on a plane, the four feet are arranged as the vertices of an equilateral triangle (of side a) and its centroid. To apply the four feet to a sphere, the middle foot is raised above the plane of the other three, the distance it rises being measured by a screw. If it rises h, where h is small compared with a, calculate the radius of the sphere.

†Ex. 265. If a frustum of a cone be such that a sphere can be inscribed in it touching both the plane ends and the curved surface, find the height in terms of the radii of the ends; and show that the circle of contact of the sphere and the curved surface of the frustum divides the spherical surface into two parts whose areas are in the ratio of the radii.

CHAPTER XII.

THE SOLID ANGLE.

At the vertex of a pyramid there is a **solid angle**. A solid angle may be defined as follows: Let O be a point outside the plane of a polygon; a straight line terminated at O and extending to infinity, when it slides round the perimeter of the polygon, encloses a region of space called a solid angle.

The solid angle is important in electricity: and here the term is widened by allowing the guide-line to be a curve instead of a polygon. A curve may, of course, be regarded as the limit of a polygon of an infinite number of infinitely short sides. There is a solid angle at the vertex of a cone.

It must be noted that 'solid angle' has nothing to do with 'angle'; a solid angle is not a kind of angle.

A solid angle has a **vertex, edges, and faces**; the faces are plane angles.

Ex. 266. What is the relation between the number of edges and of faces in a solid angle?

Ex. 267. Three planes, infinite in all directions, meet at a point. How many solid angles are formed?

Ex. 268. In what case is a solid angle fixed by its plane angles?

Solid angle and spherical polygon. Let a solid angle be cut by a sphere of unit radius and centre at the vertex of the solid angle. The faces of the solid angle are cut by the sphere in great circles, and determine a spherical polygon. The magnitude of the solid angle is measured by the area of the corresponding spherical polygon, a suitable unit being chosen.

The plane angles of the solid angle are equal to the sides of the spherical polygon (measured as angles); the inclinations of adjacent faces of the solid angle are equal to the angles of the spherical polygon.

3-faced solid angle. If a 3-faced solid angle be constructed of paper, slit along one of the edges, and **developed,** i.e. opened flat, into a plane, we shall obtain a figure such as

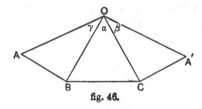

fig. 46.

In fig. 46, OA, OA' were originally coincident.

Let us now begin with a figure such as that of fig. 46, and investigate whether it can be folded into a 3-faced solid angle. We shall have to fold the planes OAB, OA'C about OB, OC respectively; and try to make OA coincide with OA'.

Now if $a > \beta + \gamma$, it will be impossible to fold OA and OA' into coincidence; even if the wings are folded over flat on to OBC, there will be a gap left between OA and OA'. If $a = \beta + \gamma$, OA and OA' will just meet when the wings are folded flat on to OBC. In order that a solid angle may be formed, it is necessary that $a < \beta + \gamma$.

From this we conclude that **any two plane angles of a 3-faced solid angle are together greater than the third.** The Euclidean proof is given below.

It is interesting to notice what this leads to when applied to the associated spherical triangle. It shews that **any two sides of a spherical triangle are together greater than the third.**

This is analogous to the theorem in plane geometry; and we may deduce that **the shortest path on a spherical surface between two points is the minor arc of the great circle joining them.**

†**Ex. 269.** Deduce from the above property of the spherical triangle the property that any side of a spherical polygon is less than the sum of the remaining sides. By passing to the limit, prove the above statement that the spherical distance is shorter than any other line on the surface of the sphere connecting two points.

The following is the Euclidean proof of the preceding theorem.

Any two plane angles of a 3-faced solid angle are together greater than the third.

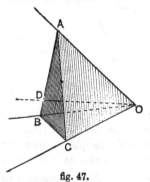

fig. 47.

Data A solid angle is bounded by three plane angles AOB, BOC, COA.

To prove that any two of these angles are together greater than the third.

Proof If all three angles are equal, any two are together greater than the third.

If they are not all equal, let ∠ AOB be that which is not less than either of the other two.

Then ∠ AOB + ∠ BOC > ∠ COA,

and ∠ AOB + ∠ COA > ∠ BOC.

It remains to prove that ∠ BOC + ∠ COA > ∠ AOB.

In plane AOB, make ∠ AOD = ∠ AOC, and OD = OC. Let AD cut OB in B and join CB.

Then in △s AOC, AOD,

OA is common, OC = OD, ∠ AOC = ∠ AOD.

∴ OC = OD and AC = AD.

Since AC + CB > AD + DB,

∴ CB > DB.

In △s BOC, BOD, OB is common, OC = OD, and CB > DB.

∴ ∠ COB > ∠ DOB.

∴ ∠ COB + ∠ COA > ∠ AOB. Q. E. D.

If the guide polygon associated with a solid angle (p. 60) is convex, the solid angle may be termed **convex**.

If the surface of a convex solid angle is slit along one edge, and developed into a plane, it is obvious that the plane angles will not completely fill up the four right angles; there will be an angular gap. Hence **the sum of the plane angles of a convex solid angle is less than four right angles.** The following is a Euclidean proof.

Let the solid angle be cut by a plane, the section being the polygon ABCDE. Take any point S inside this polygon and in its plane; join S to the vertices of the polygon.

At A there is a 3-faced solid angle; and

∠ OAE + ∠ OAB > ∠ SAE + ∠ SAB.

Similar relations hold for the solid angles at B, C, D, E.

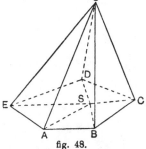

fig. 48.

Adding, we have sum of ∠ s at bases of side-faces of pyramid > sum of ∠ s at bases of △s with vertex S.

But sum of all the ∠s of side-faces = sum of all the ∠s of the S △s (the number of triangles being the same).

Hence sum of ∠s at O < sum of ∠s at S

$$< 4 \text{ right } \angle s.$$

Ex. 270. What property of a spherical polygon follows from the above theorem?

Ex. 271. How many regular solid angles are there that can be made up of angles of (i) an equilateral triangle, (ii) a square, (iii) a regular pentagon, (iv) a regular hexagon?

†**Ex. 272.** If two plane angles of a 3-faced solid angle are equal, these two faces are equally inclined to the third face.

†**Ex. 273.** The opposite edges of a certain tetrahedron are equal. The surface is slit along three concurrent edges, and opened out flat: prove that it forms an acute-angled triangle: and also that the faces are acute-angled.

†**Ex. 274.** If the sum of the angles at each of three vertices of a tetrahedron is two right angles, the same is true of the angles at the fourth vertex; and opposite edges of the tetrahedron are equal.

†**Ex. 275.** The sum of the angles subtended by the sides of a triangle ABC at any point D not in its plane is less than the sum of the angles subtended by these sides at a point P inside the tetrahedron ABCD.

CHAPTER XIII.

THE REGULAR SOLIDS. THE PRINCIPLE OF DUALITY. EULER'S THEOREM.

Ex. 276. Make a table showing in which of the following cases it is possible to make solid angles bounded by three, four, five or six plane angles when each plane angle is equal to the angle of a regular (i) 3-gon, (ii) 4-gon, (iii) 5-gon, (iv) 6-gon.

Which of these solid angles would be fixed and which could be varied in shape?

The table made in Ex. 276 shows that it is impossible to have more than five solid angles bounded by plane angles each equal to the angle of a regular polygon. Therefore it is impossible to have more than five solids all the faces of which are congruent regular polygons and all the solid angles of which are bounded by the same number of plane angles. Such solids are called **regular solids.**

Of course this does not prove that there are any regular solids; it merely shows that there is a limit to the number of them.

The figures below show the five regular solids.

Tetrahedron (i.e. 4-faced).

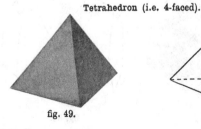

fig. 49.   fig. 50.

Cube.

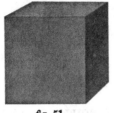

fig. 51.

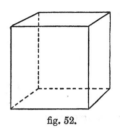

fig. 52.

Octahedron (i.e. 8-faced).

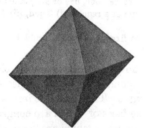

fig. 53.

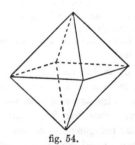

fig. 54.

Dodecahedron (i.e. 12-faced).

fig. 55.

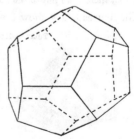

fig. 56.

Icosahedron (i.e. 20-faced).

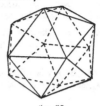

fig. 57.

fig. 58.

It is interesting to note that many substances crystallise in the form of some of the regular solids, e.g.

Form.	Substance.
Tetrahedron	Blende, Tetrahedrite (gray copper ore).
Cube	Fluor-spar, Galena (lead sulphide).
Octahedron	Alum, Diamond*, Gold, Magnetite.

One great law of crystallography is that in different crystals of the same substance the angles between corresponding faces are always the same although the faces need not be of the same size. As a consequence of this a substance which crystallises in the form of, say, a regular octahedron will

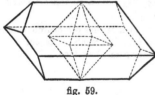

fig. 59.

fig. 60.

probably be found in the form of the figures that can be made by planing down one or more of the faces of a regular octahedron, see fig. 59. Again some substances which crystallise in the form of cubes or octahedra show combinations of these forms, e.g. fig. 60 shows a common form for crystals

---

* The natural faces are not to be confused with the artificial faces seen in cut stones. As a rule the crystals of diamond and gold are waterworn and not very well defined in shape.

of fluor-spar and galena; the triangular faces are parallel to the faces of a regular octahedron; with these two sets of faces a great variety of forms are possible according as the corner faces are well or ill developed*.

Ex. 277. Make sketches of a cube showing how to cut it so as to get sections of 3, 4, 5, or 6 sides. Is it possible to have a section of more than 6 sides?

†Ex. 278. If AB is a diagonal of a cube, prove that the mid-points of the edges which do not pass through A or B are the vertices of a regular hexagon.

Ex. 279. What numbers of sides are possible for sections of a tetrahedron? Make sketches.

Ex. 280. Calculate the length of the diagonal of a regular octahedron in terms of the length of the edges.

Ex. 281. Prove that a regular octahedron can be divided into two square pyramids in three different ways.

Ex. 282. If each corner of a cube were planed off symmetrically (see fig. 60), and the process continued till the faces of the cube disappeared, what solid figure would be constructed?

†Ex. 283. Prove that the mid-points of the faces of a cube are the vertices of a regular octahedron.

†Ex. 284. Prove that the mid-points of the edges of a regular tetrahedron are the vertices of a regular octahedron.

Ex. 285. Make sketches showing what number of sides are possible for plane sections of a regular octahedron.

†Ex. 286. Prove that the section of a regular octahedron parallel to and midway between a pair of opposite faces is a regular hexagon.

What would be the shape of any other parallel section?

THE PRINCIPLE OF DUALITY.

In plane geometry there is a certain duality by which many properties of points have, as their counterpart, corresponding properties of lines.

---

* For further information on this subject see *Crystallography* by W. J. Lewis, Cambridge University Press.

For instance:

2 points define 1 line.	2 lines define 1 point.
3 points define 3 lines.	3 lines define 3 points.

In solid geometry there is a similar duality in which points correspond to planes and lines occupy an intermediate position.

For instance:

Two points determine a line.	Two planes determine a line.
Three points determine a plane, unless they are all on the same line.	Three planes determine a point, unless they all contain the same line.
Two lines, in the same plane, determine a point.	Two lines, through the same point, determine a plane.

Ex. 287. Make a table showing the number of corners, edges and faces of the following regular solids (i) tetrahedron, (ii) cube and octahedron, (iii) dodecahedron and icosahedron.

What dual relations do you notice between the numbers of corners and faces of the solids in the above groups?

EULER'S THEOREM[*].

Ex. 288. Make a table showing the number F of faces, the number V of corners, and the number E of edges for various solids (not necessarily regular) bounded by plane faces:

Name of Solid	F	V	E	F + V − E

What inference seems probable from the table?

Imagine the vertex cut off a square pyramid, one face being thus added. By how many are the corners increased? and the edges? Give the value of $F + V - E$ for this new solid. Also suppose a different corner of the square pyramid cut away, and see what effect this has on the number $F + V - E$.

---

[*] Euler was born at Basle 1707, and died at St Petersburg 1783. He published this theorem in 1758, but it seems to have been worked out previously by Descartes.

Ex. 289. Consider a hollow box in the shape of a solid bounded by plane faces.

If we cut out one face of the solid, we do not abolish any edge or any vertex, but we are left with a number of *raw* edges, edges that now belong to one face only instead of being common to two faces.

Now let the box be further dismembered face by face, subject to two conditions (i) that no face is to be removed that has not at least one raw edge, (ii) that no face is to be removed whose raw edges are not consecutive.

Make a table, as below:

Number of face removed	Number of edges abolished	Number of vertices abolished
1st		
2nd		

Ex. 290. Take a hollow tetrahedron to pieces and make a table as above.

Ex. 291. Repeat Ex. 290 for (i) a hexagonal prism, (ii) a square pyramid, (iii) an octahedron.

Euler's Theorem. If E, V, F are the number of edges, vertices, and faces of a solid bounded by plane faces, in general $F + V = E + 2$.

Proof. Imagine a hollow solid bounded by plane faces to be taken to pieces face by face, subject to the two conditions (i) that, with the exception of the first face, every face removed has at least one *raw* edge (i.e. an edge belonging to one face only instead of being common to two faces); (ii) that no face is to be removed whose raw edges are not consecutive.

After any number of faces have been taken away suppose that there is a face which has x consecutive raw edges; then the removal of this face will wipe out x edges but only $\overline{x-1}$ vertices.

Fig. 61 will make this clear. The thin lines represent raw edge, so that the removal of this face will wipe out 4 edges but only 3 corners.

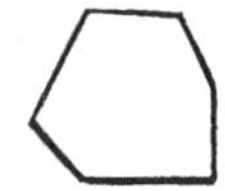

Let e_r, v_r denote the number of edges and vertices wiped out by the removal of the rth face.

fig. 61.

Then $e_r - v_r = 1$ except for the first and the last face.

Now the removal of the first face destroys no edges or vertices.

$$\therefore \ e_1 - v_1 = 0.$$

And the removal of the last face destroys the same number of edges as vertices.

$$\therefore \ e_n - v_n = 0.$$

$\therefore$ if E, V, F are the total number of edges, corners and faces,

$$E - V = F - 2,$$

or $$F + V = E + 2.$$

This result applies, in general, to any solid bounded by plane faces, but no account has been taken here of a solid any face of which has more than a boundary, for example a solid such as that formed by placing a cuboid on a larger cuboid.

Ex. 292. What is the value of $F + V - E$ for the solid mentioned above?

Ex. 293. Verify Euler's Theorem for each of the regular solids.

CHAPTER XIV.

COORDINATES IN SPACE

Ex. 294. Suppose the walls of a room run due N. and S. and due E. and W. Do the following directions send you to a definite point? From the N.W. corner of the floor go 3 feet due E., then 4 feet due S., then 5 feet vertically upwards.

Ex. 295. Explain what measurements you would have to make to give directions to go from the N.W. corner of the floor to some point in the room, e.g. the bottom of an electric light bulb, the corner of a desk.

In a plane you have seen that if two fixed perpendicular axes are chosen the position of any point in the plane is determined by two coordinates. In space we have to take three perpendicular axes and the position of any point is determined by giving the distances we must travel from the origin parallel to each of these three axes. (These distances are of course the same as the distances of the point from the three planes containing the axes in pairs.)

It is usual to place these axes as in fig. 62, OX, OY, OZ being their positive directions.

Note that the three planes correspond to the floor and walls of the room in Ex. 294, O corresponding to the N.W. corner of the floor.

Ex. 296. Find the length of OP, P being the point (4, 2, 4); see fig. 62.

Ex. 297. Make a sketch like fig. 62, showing the points (5, −2, 3), (−6, 4, −1). What are their distances from O?

Ex. 298. Find the distances between the points (i) (1, 2, 3) and (8, 5, 7), (ii) (a, b, c) and (p, q, r).

Ex. 299. If P is the point (4, 2, 4), find the angles OP makes with the planes XOY, XOZ, YOZ; also find the angles OP makes with the axes.

Ex. 300. What is the length of the projection (i) upon the x axis, (ii) upon the y, z plane of the line joining (a, b, c) to (x, y, z)?

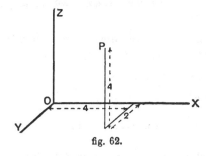

fig. 62.

In plane geometry, if you are confined to points whose coordinates satisfy one equation in x and y, your locus is a line (straight or curved).

In geometry of three dimensions, if you are confined to points whose coordinates satisfy one equation in x, y, z; e.g.

$$z = 5x^2 + 3xy + 2y - 5,$$

you may choose any values you please for x and y and there is a corresponding value for z. You see that the restriction imposed by one equation compels you to move on a surface of some sort; the equation is called the equation of the surface.

Ex. 301. What restriction is imposed by two equations (i) in plane geometry, (ii) in solid geometry?

Ex. 302. What restriction is imposed by three equations in solid geometry?

Ex. 303. If you are told to go 2 units in the direction OX and then given absolute freedom to move any distances parallel to OY, OZ, i.e. if the only condition imposed on you is $x=2$, what is your locus?

Ex. 304. What is the locus $x=2$ in plane geometry?

Ex. 305. What is the locus $y=-3$ in solid geometry? What is the locus $z=0$ in solid geometry?

Ex. 306. What are the loci $x=y$, $x=z$ in solid geometry?

Ex. 307. What is the locus $x^2+y^2=a^2$ (i) in plane geometry, (ii) in solid geometry?

Ex. 308. What is the locus $ax+by+c=0$, (i) in plane geometry, (ii) in solid geometry?

It can be shown that the locus corresponding to an equation of the first degree in x, y, z is a plane surface.

Ex. 309. What are the equations of the planes XOY, YOZ, ZOX?

Ex. 310. Find the equation of a plane which cuts off lengths a, b, c from the axes.

Ex. 311. What locus is determined by two equations of the first degree?

Ex. 312. What are the equations of OX, OY, OZ?

Ex. 313. What is the distance of the point (x, y, z) from the origin? Hence find the equation of a sphere of radius r whose centre is at the origin.

Ex. 314. What is the equation of a sphere of radius r whose centre is at the point (a, b, c)?

Ex. 315. A cord is stretched in a room between two points in space, and the three planes of reference are (i) the north wall of the room, (ii) the east wall, and (iii) the floor. If the coordinates of the two points between which the cord is stretched tightly are $(5, 2, 4\frac{1}{2})$ and $(4, 5\frac{1}{2}, 1\frac{1}{2})$, unit 1 foot, calculate the length in feet of this string, and the points on the floor and east wall where the cord, if prolonged, would intersect them. What angle does the string make with the floor?

CHAPTER XV.

PLAN AND ELEVATION.

Ex. 816. If the projection of a point P on a given plane is fixed, what is the locus of P?

It is obvious that, if the projection of a point on a given plane is given, and also the distance of the point from the plane, the position of the point is fixed. A plane figure showing the projections on the plane of a series of points and the distances of the points from the plane is called an **indexed plan**: e.g. fig. 63 is the indexed plan of a tetrahedron with its base parallel to the plane of projection.

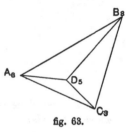

fig. 63.

It will be noticed that a contour map is an indexed plan.

The position of a point may also be fixed by giving its projections on two planes. It is usual to take one plane horizontal and the other vertical; the projection on the horizontal plane is called the **plan** and the projection on the vertical plane the **elevation**. The horizontal plane of projection is spoken of as the H.P. and the vertical plane as the V.P.

The line of intersection of the H.P. and V.P. is called the **ground line** and is denoted by ˙XY.

Fig. 64 shows a cube; its plan is the square on which it stands, its elevation is shown on the vertical plane. Note that a, a' are the plan and elevation of A, and that when any point in, say the plan, is the plan of two points it is lettered as a fraction, thus $\dfrac{a}{p}$, A being the point nearer to the spectator, and P the point nearer to the plane of projection.

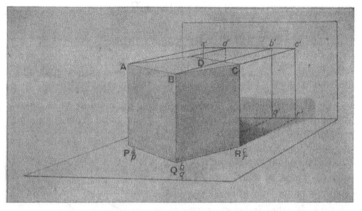

fig. 64.

If the vertical plane is rotated about XY through an angle of 90° we have the plan and elevation in the same plane. For convenience the plan and elevation are generally drawn on the same paper; thus fig. 65 represents the plan and elevation of the cube in fig. 64; broken lines are drawn connecting the plan and elevation of each point.

Ex. 317. What is the angle between XY and the line joining the plan and elevation of any point?

Ex. 318. Sketch* the plan and elevation of (i) a vertical line, (ii) a horizontal line, (iii) a sloping line with one end in the v.p.

---

* In sketching plans and elevations it is useful to fold a sheet of paper at right angles to represent the H.P. and V.P. and to place the model in position.

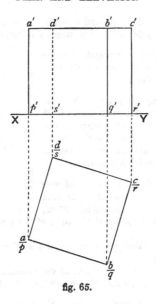

fig. 65.

Note that the elevation of a point G below the H.P. is a point g' below the XY line; g' must not be confused with a point in the plan. In the same way the plan of a point H behind the V.P. is a point h above the XY line; h must not be confused with a point in the elevation. In general we shall only consider points above the H.P. and in front of the V.P.; it is obvious that in most cases this does not cause any loss of generality, for the H.P. and V.P. can be chosen to be respectively below and behind all the parts of the solid we wish to represent.

Ex. 319. Draw a figure showing the plan and elevation of (i) a point P which is on the H.P. and behind the V.P., (ii) a point Q above the H.P. and behind the V.P., (iii) a point R below the H.P. and in front of the V.P., (iv) a point S below the H.P. and behind the V.P.

Ex. 320. Draw the plan and elevation of a square pyramid with its base on the H.P., the corner of the base nearest the V.P. being 1 in. from the ground line, the nearest edge of the base making an angle of 30° with the ground line, length of edge of the base 2 in., height of pyramid 2 inches.

Ex. 321. Sketch the plan and elevation of several solids in various positions, e.g. a triangular prism, a pentagonal prism, a tetrahedron.

Ex. 322. Given the plan and elevation of a solid, show how to draw an elevation on another vertical plane.

Ex. 323. In fig. 66 you are given the projections of a piece of wire AB fixed in the H.P. at B. From the plan project a new elevation of the wire on the vertical plane through X′Y′, and measure this elevation. Dimensions are given in centimetres.

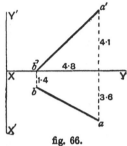

fig. 66.

Ex. 324. A regular square pyramid rests with a triangular face on the H.P. The plan *abv* of this face is a triangle having *ab* = 2·2 in., *va* = *vb* = 3 in. Draw the plan and elevation of the pyramid on a V.P. perpendicular to *ab*.

[First draw △ *abv*, then the elevation of a section by a plane through *v* parallel to the V.P.]

Also find the shape of the section of the pyramid by a vertical plane cutting *va*, *vb* at points *q*, *r* such that *vq* = 0·9 in., *vr* = 1·3 in.

Ex. 325. ABC and DEF are thin 60° and 45° set-squares; angles A and D are the right angles and ∠C = 60°, AB = EF = 3 in. Imagine ABC, DEF placed flat on the H.P. so that B is at the mid-point of EF and EF parallel to CA. Now suppose ABC rotated about AB till its plane is vertical, and DEF rotated about EF so as to rest against the edge of the other set-square. Draw the plan of the set-squares and elevations of them on the vertical planes parallel to AB and EF.

Fig. 67 shows a perspective sketch of a cottage, its plan, and its elevations on two vertical planes at right angles to one another. These figures are reduced from an architect's actual drawings; details of measurements, etc. are omitted for the sake of clearness*.

---

* In an architect's drawings the side elevation would be placed upright by the front elevation.

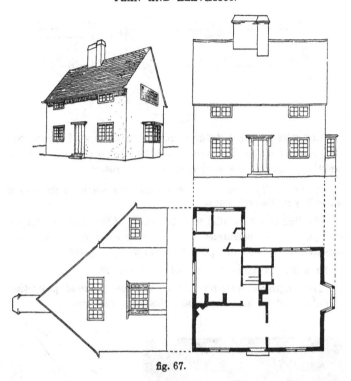

fig. 67.

Note that the line joining any point on the plan to the corresponding point on the elevation cuts the ground line at right angles. Also note that the distance of any point from the ground line in the front elevation is equal to the distance of the corresponding point from the ground line in the side elevation.

Ex. **326.** Fig. 68 shows the plan and end elevation of the roof of a house; make a sketch of the side elevation.

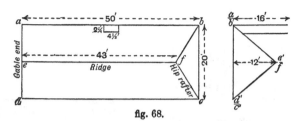

fig. 68.

Lengths of lines determined from plan and elevation.
Of course if a line is parallel to either plane of projection its
length is equal to its projection on that plane.

Ex. 327. What lines in fig. 68 are the same lengths as the lines of
which they are the projections?

If a line is not parallel to either plane of projection its length
can generally be calculated by measuring lines in the plan and
elevation figure and using the theorem of Pythagoras.

Ex. 328. Determine the length of FB from fig. 68.

The lengths of lines not parallel to either plane of projection
can also be determined by geometrical constructions.

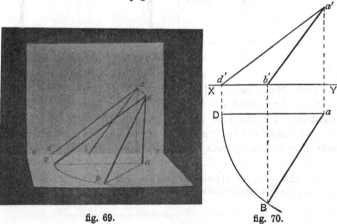

fig. 69.   fig. 70.

Fig. 69 shows a line AB in space and fig. 70 its plan and elevation. If the triangle ABa is rotated about Aa to the position ADa parallel to the vertical plane, $a'd'$, the elevation of AD, will be of the same length as AD or AB; this suggests a construction for determining the length of the line of which Ba, $b'a'$ (fig. 70) are the plan and elevation. What path will B in fig. 69 describe as ABa swings round Aa?

*Ex. 329. Determine the length of the line PQ using the dimensions given in fig. 71.

*Ex. 330. Determine geometrically the length of FB in fig. 68.

Ex. 331. A stick rests with one end on a floor, a metre from the north wall of the room and a metre from the west wall. The other end rests on the west wall at a height of 1·4 metres and 0·8 metres from the north wall.

Show the projections of the stick on the floor and west wall, and find the length of the stick.

The upper end of the stick slips, the lower end keeping still. Show the path of the upper end.

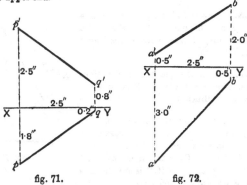

fig. 71.   fig. 72.

*Ex. 332. Fig. 72 represents the plan and elevation of a line AB. If the line is produced to cut the H.P. at h, then h', the elevation of h, must lie on $b'a'$ produced. What other line is there in the figure on which h' must lie? Hence find the plan of h. Measure the distance of h from XY.

---

* Numerical values are given in these Exs., so that an accurate drawing may be made if desired; in cases in which the technical skill is unimportant, rough sketches will be sufficient.

*Ex. 333. Explain how to find the plan and elevation of the point v' at which AB (see fig. 72) cuts the V.P. Measure the distance of v' from XY.

The points at which a line cuts the H.P. and V.P. are called the **horizontal trace** (H.T.) and the **vertical trace** (V.T.) of the **line.**

Ex. 334. Suppose h, and v', the H.T. and V.T. of a line are given, show how to draw the plan and elevation of the line.

Ex. 335. Having given the traces of a line and the projections of a point, what conditions must obtain if the point is on the line?

Ex. 336. Explain how to use the theorem of Pythagoras to find the length of a line whose ends are in the V.P. and H.P., when its traces are given.

*Ex. 337. Find the H.T. and V.T. of the lines whose plan and elevation are given in (i) fig. 71, (ii) fig. 73.

Ex. 338. Draw the plan and elevation of a line. If p, p' are the plan and elevation of a point P on the line, what geometrical conditions must obtain?

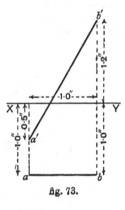

fig. 73.

Ex. 339. Given the plans and elevations of 8 points, what are the conditions that the three points are collinear?

Ex. 340. Mark on a plane points h, $(0, -a)$; v', (b, c); p, $(d, -e)$; p', (d, f). Suppose the x axis to be the ground line, p, p' the plan and elevation of a point P; what relations must exist between a, b, c, d, e, f if P is on the line of which h, v' are the H.T. and V.T.?

Ex. 341. On the V.P. draw a pair of intersecting lines; also draw a pair on the H.P. If these lines represent the plan and elevation of two straight lines, what is the condition that those lines should intersect?

---

* Numerical values are given in these Exs., so that an accurate drawing may be made if desired; in cases in which the technical skill is unimportant, rough sketches will be sufficient.

Ex. 842. Given the traces of two straight lines, explain how to determine whether the lines intersect.

Inclination of a straight line to the H.P. and V.P.

On p. 17 the inclination of a straight line to a plane has been explained to be the angle between the line and its projection on the plane.

*Ex. 843.** Find the inclination to the H.P. of the following lines in fig. 68, AE, FB.

Ex. 844. Show how to determine the angle made with (i) the H.P., (ii) the V.P. by a line whose ends are in the planes of projection and whose plan and elevation are given. [See figs. 69, 70.]

Note that if the ends of a line are not in the planes of projection, the line can be moved parallel to itself till its ends come into those planes, so that the general problem of the inclination of a line to the H.P. and V.P. can easily be reduced to the special case of Ex. 344 (see fig. 74. PG, RS are parallel and equal).

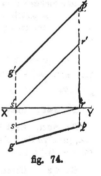

The method suggested by fig. 69 of rotating a plane about its line of intersection with the V.P. (or about a line parallel to the line of intersection) until the plane coincides with (or is parallel to) the V.P. is much used

fig. 74.

for determining the true shape of a plane figure; the figure obtained is called the **vertical rabattement**. If a figure is rotated into a horizontal position, the figure obtained is called the **horizontal rabattement**.

---

* Ex. 343 can be easily done by Trigonometry, but the geometrical constructions are interesting and in practical use.

*Ex. 845. A thin 45° set-square, hypotenuse AB=4 in., has the edge AC in the v.p. at right angles to XY, and the edge CB in the h.p. and inclined at 50° to XY.

(i) Draw the plan and elevation of the set-square.

(ii) Draw the horizontal rabattement of △ABb', and hence find the angles which AB makes with the v.p.

(iii) Find the angle between AB and XY.

*Ex. 846. A straight line 3 in. long has its ends in the h.p. and v.p. respectively and makes an angle of 35° with the h.p. and lies in a vertical plane which makes an angle of 72° with the v.p. Draw the projections of the line and measure their lengths.

*Ex. 847. Draw the projections of a line 7 cm. long which has its ends in the h.p. and v.p. respectively and which makes angles 20°, 57° with those planes. Measure its projections.

*Ex. 848. Show how to draw the projections of a line which has its ends in the v.p. and h.p. respectively, the length of the line being 3·5 in., and the lengths of its plan and elevation 2·8 in., 2·3 in. Measure the angles the line makes with the v.p. and h.p.

[Make a sketch of the solid figure; suppose the ends of the line to be v', h, draw the triangle obtained by swinging △$v'hv$ about vv', draw the locus of h as the triangle swings round.]

In the same way the lines in which a plane cuts the h.p. and v.p. are called the horizontal and vertical traces of the plane.

Ex. 849. Sketch the traces of any plane; where do they intersect?

Ex. 850. Sketch the traces of (i) a vertical plane perpendicular to the v.p., (ii) a vertical plane inclined to the v.p., (iii) a vertical plane parallel to the v.p., (iv) a horizontal plane.

Ex. 851. Given the traces of a plane and of a line on it, what conditions obtain?

Ex. 852. Given the traces of a plane and the plan and elevation of a line, show how to determine whether the line is in the plane.

[Find the traces of the line.]

---

* Numerical values are given in these Exs., so that an accurate drawing may be made if desired; in cases in which the technical skill is unimportant, rough sketches will be sufficient.

Ex. 858. Given the traces of a plane and the plan and elevation of a point, show how to determine whether the point is in the plane.

[Consider the line joining the point to a point on one of the traces.]

Ex. 854. Having given the traces of a plane, determine the inclination of the plane to the H.P.

[Draw X'Y' perpendicular to the H.T.; on X'Y' as ground line draw an elevation.]

Given the plan and elevation of a plane figure, to find its true shape.

The method used is (1) to find the line of intersection of the plane of the figure with the H.P. (i.e. the H.T. of the plane), (2) to imagine the figure rotated about this line into the H.P. or a vertical plane, i.e. to find the horizontal or a vertical rabattement of the figure.

Before considering the problem we will take an example of the converse problem.

Ex. 855. A circle of radius 2 in. lies in a vertical plane which makes an angle of 60° with the V.P. its centre being in the H.P., draw its elevation.

See fig. 75. *aPb* represents the circle rotated about the horizontal diameter into the H.P. *Pp* is of course equal to the height of *p'* above XY.

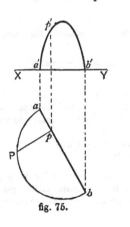

fig. 75.

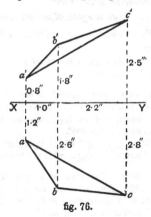

fig. 76.

Ex. 356. Find by the following construction the true shape of the triangle whose plan and elevation are given in fig. 76.

(i) Draw fig. 76 accurately and find the horizontal traces of the sides of △ABC. The horizontal traces should lie on a straight line (X'Y'); give a reason for this.

(ii) Draw an elevation of △ABC on the vertical plane at right angles to X'Y'. Explain the figure you obtain.

(iii) Now suppose the plane of the triangle rotated about X'Y' until it coincides with the H.P.; draw the new plan. Measure the sides of true triangle.

[The figure you obtained in (ii) will give you the distances of A, B, C from X'Y'.]

Ex. 357. Draw the plan and elevation of a circle of radius 2 inches whose horizontal diameter makes an angle of 50° with the V.P. and whose plane is inclined at 35° to the H.P.

MISCELLANEOUS EXAMPLES ON PLAN AND ELEVATION.

Ex. 358. A thin 30°—60° set square ABC, whose long edge AB measures 4 inches, is laid on the horizontal plane and is then turned about AB through an angle of 58°;

(i) Draw the plan of the set square;

(ii) Determine the inclinations to the horizontal of the edges AC and BC;

(iii) Draw an elevation of the set square on a vertical plane which makes 45° with AB.

Ex. 359. The plan of a roof of a house is rectangular in outline, and two adjacent surfaces are each inclined at 50° to the horizontal. Find the inclination of the hip rafter (see fig. 68) to the horizontal and to the bottom edges of the roof.

What is the length of the hip rafter, if 20 feet is the length of the bottom edge of one of the slopes which is triangular in shape?

Ex. 360. The plan of a roof of a house is rectangular in outline (see fig. 68); the side slopes are inclined to the horizontal at 30°, the end slope at 40° and the span is 24 feet. Find the length of the hip rafter and its inclination to the horizontal.

Ex. 361. Fig. 77 is the dimensioned roof plan of a house. The roof planes A and B are inclined to the horizontal at 33°. Determine

(i) The length and inclination of the valley rafter CD;

(ii) The bevel for cutting the edges of the slates which lie along the valley, that is, the true angle between the lines whose plans are *cd*, *de*;

(iii) The correct angle of the valley tiles, that is, the dihedral angle between the planes A and B.

Ex. 362. If the roof in fig. 77 were hipped back at E to the same slope as that of the planes A and B, would the area to be slated be increased, decreased or unaltered?

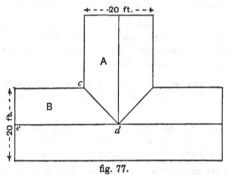

fig. 77.

Ex. 363. A thin triangular plate ABC ($\angle$ C = 90°, BC = 2½", CA = 4") is bent about the bisector of the angle ABC till the two parts are at right angles. Find by drawing (i) the distance from A to C, (ii) the angle BA makes with BC.

Ex. 364. An eastward path on the plane face of an embankment rises 10 feet vertically in a horizontal distance of 15 feet, and a southward path falls 10 feet vertically in a horizontal distance of 20 feet. Find the direction of the path of steepest ascent, and the inclination, in degrees, of this path to the horizontal.

Ex. 365. Do Ex. 315 by drawing.

Ex. 366. Having given a plane by its traces, and the plan of a point A on the plane, show how to find the traces of a line of steepest slope through A.

Ex. 367. A triangular prism, 2 in. long, with equilateral ends of 2·5 in. side, rests with a long edge on the ground, the end faces being inclined at 50° to the horizontal. Draw its plan. Determine the section made by a horizontal plane 1·5 in. high.

Ex. 368. The tripod that supports a photographer's camera has three legs 130 cm. long meeting at a point. Show that when the tripod stands on level ground, no matter how it may be placed, the legs are equally inclined to the ground. When the feet form an equilateral triangle 70 cm. in the side, calculate the height of the tripod.

With rectangular axes, z-axis vertical, unit 1 cm., the feet of the tripod are at (33, 42, 0), (97, 20, 0) and (68, 114, 0). On a scale of $\frac{1}{10}$ show the tripod in elevation on the plane $x = 0$, and in plan.

Ex. 369. Through a point 3 inches above the H.P. pass two lines including an angle of 40°; they are inclined at 50° and 65° respectively to the H.P. Find, by drawing plan and elevations, the inclination to the horizontal of the plane containing these lines.

Ex. 370. Determine the H.T. and V.T. of a plane which bisects the line PQ of fig. 71 at right angles. Also find the true angle between the traces of this plane.

Ex. 371. The H.T. and V.T. of a plane make 60° and 45° with the ground line. A point p, distant 1″ from the H.T. and $1\frac{1}{2}$″ from XY, is the plan of a point P contained by the plane; draw the elevation of P.

Also draw the projections of a line PM which is perpendicular to the plane and has its foot M in the H.P.

Ex. 372. Three balls, 5 cm. in diameter, lie on a floor in contact, and a fourth equal ball is placed on them. Draw them in plan and in elevation. State how you determine the height of the centre of the fourth ball above the plane of the other three centres.

CHAPTER XVI.

PERSPECTIVE.

It is not part of the plan of this book to deal with the practice of perspective drawing; the principles, however, are a simple and interesting application of solid geometry and, as such, are worthy of the reader's attention. Here we shall limit ourselves to an explanation of the simplest points of the theory.

The first problem met with in drawing a picture is to represent on a plane two-dimensional surface objects that are in three dimensions. The third dimension—depth—cannot exist on the plane surface of the picture and must be suggested to the spectator by various devices. One of these devices is to give the idea of distance by representing distant objects as more or less hazy, an intervening atmosphere being suggested by this means. The necessary illusion, however, is obtained mainly by the use of **geometrical perspective**, by which distant objects are depicted as *smaller* than near objects, on a systematic plan which shall now be explained.

Suppose that it is required to depict a three-dimensional object (say a cube) on a pane of glass placed between the eye and the cube. Suppose that the eye is fixed by compelling it to look through a fixed eye-piece (e.g. a small hole in a fixed piece of paper or cardboard).

If the glass is within arm's length, it will be possible to trace upon it an ink outline exactly covering the outlines of the cube. A picture is obtained by these means which will be found, if looked at from any point of view, to suggest the form and position of the cube: though the illusion is not complete except when the eye returns to its original position relative to the picture.

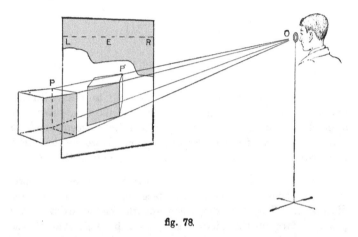

fig. 78.

It will be noticed that the relative dimensions on the picture do not correspond to the relative dimensions in the object; e.g. the pictures of equal edges are not equal, if the edges are at different distances from the eye. Again, parallel lines of the cube will probably be depicted by lines that are not parallel.

Let us now express in geometrical language the process that has taken place in drawing this picture. Let O be the point at which the eye was placed. The pane of glass we will call the **picture plane,** and we will suppose it to be vertical. A horizontal plane through O will cut the picture plane in a horizontal line LR, which we will call the **horizon line.** The line through O perpendicular to the picture plane will be called the **line of sight**; the point, E, where the line of sight cuts the picture plane is the **centre of the picture**; this point lies on the horizon line.

Let the line joining O to any point, P, of the object cut the picture plane in a point P'; P' is the picture of P, or, in mathematical language, the **projection*** of P. OP we will call the **projecting line of P**. The picture is, in fact, the projection of the object, from O, upon the picture plane.

Projection in this wider sense is a generalisation of the orthogonal projection already familiar to the reader. Orthogonal projection is a particular case of projection; for if the centre of projection (the eye) is removed to infinity, the projecting lines all become parallel, and if the picture plane is now taken at right angles to the projecting lines, we have orthogonal projection.

If we consider the points of a line in the object, the projecting lines of these points form a plane, which may be called the **projecting plane** of the line. This cuts the picture plane in a straight line, the projection of the straight line depicted.

A set of parallel lines in the object (not necessarily coplanar) will determine a set of projecting planes through O; and the line through O parallel to the set will lie in each of these planes. The planes will, in fact, be arranged like the leaves of a partly opened book.

Such a set of planes will be cut by the picture plane in a set of lines, either parallel or concurrent. The section will consist of parallel lines if the picture plane is parallel to the back of the book; this will happen if the parallel lines depicted are parallel to the picture plane. Hence **parallel lines are depicted as parallel if they are also parallel to the picture plane.**

---

* Sometimes called **conical** projection (the projecting lines forming cones) to distinguish it from orthogonal projection.

A set of vertical lines is always parallel to a vertical picture plane; they will be depicted as parallel; in fact their projections will be perpendicular to the horizon line.

A set of horizontal parallel lines is not necessarily parallel to the picture plane; and will not necessarily appear as a parallel set in the picture.

If the set of parallels is not parallel to the picture plane, the latter will cut the 'book' of planes in a set of concurrent lines. Their point of concurrence is the point where the picture plane cuts the back of the book; in other words, where the line through the eye parallel to the set cuts the picture plane. This is called the **vanishing point** of the set; the parallel lines will appear as a set of lines converging on this vanishing point. This point is the projection of the point at infinity on the set of parallels.

The depiction of a set of parallels by means of converging lines corresponds to our every-day experience of the appearance of such lines. If a straight length of railway is viewed from a bridge, the parallel rails appear to converge. Again, clouds will often be seen to converge upon two opposite points of the horizon; say S.W. and N.E. This is a perspective effect; in reality such clouds are arranged in parallel bands stretching from S.W. to N.E.

As already stated, the vanishing point for a set of parallels is the point where the parallel through the eye meets the picture plane. If the parallels are also horizontal (as in the roof-lines and ground-lines of a building), the vanishing point will clearly lie on the horizon line. If the parallels are not horizontal, a little consideration will shew whether the vanishing point is above or below the horizon line. If the parallels are parallel to the picture plane, the parallel through the eye cuts the picture plane at infinity; and the parallels will be depicted by parallels, as already shewn.

These considerations will enable the reader to draw a rough perspective sketch of a simple object.

MISCELLANEOUS EXERCISES.

Ex. 373. AB, AC are two straight lines which meet at an angle of 45°, AB is 10 cms. in length; at B, BD is drawn perpendicular to the plane BAC and 2·5 cms. in length. Calculate the length of the perpendicular from D to AC.

Ex. 374. ABCD is a rectangle whose diagonals meet in E, the length of each diagonal being 1·5″. Through E a straight line EF, 1·8″ in length, is drawn perpendicular to the plane of the rectangle. Find the centre of the sphere passing through ABCDF, and calculate the length of its radius to two decimal places.

†**Ex. 375.** ABCD is a tetrahedron, such that the bisectors of the angles BAC, BDC meet at a point in BC; find the relation between the edges AB, AC, BD, CD.

Prove that the bisectors of the angles ABD, ACD meet at a point in AD.

†**Ex. 376.** A, B, C, D are any four points in space; prove that the straight line which joins the mid-point of AB to the mid-point of CD intersects the straight line which joins the mid-point of AC to the mid-point of BD, and that both lines are bisected at their point of intersection.

Shew that the straight line joining the mid-points of BC and AD is also bisected at that point.

†**Ex. 377.** ABCD is a regular tetrahedron, and, from the vertex A, a perpendicular is drawn to the base BCD, meeting it in O; shew that three times the square on AO is equal to twice the square on AB.

†**Ex. 378.** Prove that in a tetrahedron ABCD, the sum of the angles ABC, ADC, BAD, BCD is less than four right angles.

Ex. 379. A sphere is inscribed in a cube, and a plane passing through the other extremities of the three edges which meet in one angle of the cube cuts the sphere in a circle, about which a square is described. Prove that the area of this square is two-thirds of the area of any face of the cube.

†**Ex. 380.** From a point E draw EC, ED perpendicular to two planes CAB, DAB, which intersect in AB, and from D draw DF perpendicular to the plane CAB, meeting it in F; shew that the line joining the points C and F is perpendicular to AB.

†Ex. 381. If BCD be the common base of two pyramids, whose vertices A and A' lie in a plane passing through BC, and if the two lines AB, AC be respectively perpendicular to the faces BA'D, CA'D, prove that one of the angles at A, together with the angles at A', make up four right angles.

†Ex. 382. Within the area of a given triangle is described a triangle, the sides of which are parallel to those of the given one; prove that the sum of the angles, subtended by the sides of the interior triangle at any point not in the plane of the triangles, is less than the sum of the angles subtended at the same point by the sides of the exterior triangle.

†Ex. 383. If O be a point within a tetrahedron ABCD, prove that the three angles of the solid angle, subtended by BCD at O, are together greater than the three angles of the solid angle at A.

†Ex. 384. From the extremities of the two parallel straight lines AB, CD parallel lines Aa, Bb, Cc, Dd are drawn, meeting a plane in a, b, c, d; prove that AB is to CD as ab is to cd, taking the case in which A, B, C, D are on the same side of the plane.

†Ex. 385. Shew that the perpendicular dropped from the vertex of a regular tetrahedron upon the opposite base is treble of that dropped from its own foot upon any of the other bases.

†Ex. 386. A triangular pyramid stands on an equilateral base, and the angles at the vertex are right angles; shew that the sum of the perpendiculars on the faces, from any point of the base, is constant.

†Ex. 387. Two planes intersect; shew that the loci of the points, from which perpendiculars on the planes are equal to a given straight line, are straight lines; and that four planes may be drawn, each passing through two of these lines, such that the perpendiculars from any point in the line of intersection of the given planes, upon any one of the four planes, shall be equal to the given line.

†Ex. 388. Three straight lines, not in the same plane, intersect in a point, and through their point of intersection another straight line is drawn within the solid angle formed by them; prove that the angles which this straight line makes with the first three are together less than the sum, but greater than half the sum of the angles which the first three make with each other.

†Ex. 389. If three straight lines which do not all lie in one plane be cut in the same ratio by three planes, two of which are parallel, shew that the third will be parallel to the other two, if its intersections with the three straight lines are not all in one straight line.

†Ex. 390. Two regular pyramids are described, the one standing on a square as a base, the other on a regular octagon; if the slant-edges of the two pyramids be equal, and the perimeters of the bases be equal, prove that the plane angles at the vertex of the former are together greater than the plane angles at the vertex of the latter.

†Ex. 391. Prove that there are two points equidistant from the plane of an acute-angled triangle at which each of the sides subtends a right angle.

†Ex. 392. If a cube and an octahedron have a common circumscribed sphere, shew that their surfaces are in the same ratio as their volumes.

†Ex. 393. Upon a diameter AOB of a circle whose centre is O two points C and D are taken such that $OC \cdot OD = OA^2$; and upon CD as diameter a second circle is described, its plane being perpendicular to that of the first circle. Prove that, if P and Q are any two points on the first circle, and E and F any two points on the second, PE : PF = QE : QF.

†Ex. 394. Find in a given line the point P which is such that $AP + PB$ is least, where A and B are two fixed points not in one plane with the line.

†Ex. 395. At the ends A, B of a chord of a sphere are drawn the tangent planes, and through any point P of the sphere are drawn the line LPM parallel to AB to meet the planes in L and M, and the line PN parallel to either plane to meet AB in N. Prove that $LP \cdot PM = PN^2$.

†Ex. 396. A perpendicular ON is drawn from a given point O to a given plane, and a variable line OP meets the plane in P. The shortest distance between the line OP and a given line parallel to ON is QR. Shew that if the product $NP \cdot QR$ is constant the locus of P consists of two parallel straight lines.

Ex. 397. AB, BC, CD are edges of a cube, of which AD is a diagonal. Prove that the angle between the planes ABD, ACD is equal to two-thirds of a right angle.

†Ex. 398. Four equilateral triangles are arranged so that their vertices coincide and their bases form a square. Shew that the opposite edges of the solid angle at the vertex are at right angles.

†Ex. 399. A straight line of constant length moves in space so that it subtends a right angle at each of two fixed points; find the locus of its middle point.

†Ex. 400. Prove that one line and only one drawn through any point in space meets at finite distances two given straight lines which are not coplanar, except when the point lies in one of two planes.

†Ex. **401.** Two circles are drawn in different planes; shew that there is in general one point on the line of intersection of the planes the tangents from which to the circles are equal in length; and that, if there be more than one, the circles are sections of the same sphere.

†Ex. **402.** The common perpendicular of two straight lines APP' and BQQ' is AB, also M and M' are the middle points of PQ and P'Q' respectively. Prove that either the common perpendicular of MM' and AB bisects AB or MM' bisects AB.

Ex. **403.** Shew that the diagonal of a pentagonal face of a regular dodecahedron inscribed in a sphere is equal to the side of the cube inscribed in the same sphere.

†Ex. **404.** ABCO is a tetrahedron such that AO is perpendicular to the plane BOC. HG is the shortest distance between OB and AC, and CD is the perpendicular from C upon OB. Prove that $CG : GA = CD^2 : AO^2$.

†Ex. **405.** If ABC, A'B'C' be two straight lines in space which do not meet, such that O, O', O'', the middle points of AA', BB', CC' are collinear, then the middle points of OA, O'B, O''C are also collinear.

†Ex. **406.** AB, AC, AD are three edges of a cube, and they are produced to E, F, G, so that the lines ABE, ACF, ADG are equal; shew that, within certain limits, the plane EFG cuts the cube so as to form a hexagon, whose alternate sides are equal, and each angle four-thirds of a right angle.

Ex. **407.** A seam of coal has its line of greatest slope running down due north and sloping downwards at 20° with the horizontal. Long ago the earth cracked along a line running N.W. and S.E., and the part N.E. of this crack rose 80 feet. Then natural forces planed the earth away till the surface was level. Draw a plan on a scale of 1 inch to 100 feet, shewing where the crack and the two parts of the seam appear at the surface.

†Ex. **408.** Points A', B', C' are taken on the edges VA, VB, VC of a tetrahedron VABC. If AB meets A'B' in C'', and BC meets B'C' in A'', and CA meets C'A' in B'', prove that A'', B'', C'' lie on a straight line.

†Ex. **409.** If the three angles at the vertex of a tetrahedron are right angles, shew that the line joining the vertex to the orthocentre of the base is perpendicular to the base.

†Ex. **410.** The square of the distance between two points is equal to the sum of the squares of its projections on three straight lines at right angles to one another.

†Ex. **411.** Shew how to construct a parallelepiped having two given finite skew lines as two edges.

†**Ex. 412.** Given two similar polygons, in different planes, with corresponding sides parallel, prove that the polygons are sections of a certain pyramid.

†**Ex. 413.** Shew how to determine a line to be cut by three given skew lines in a fixed ratio.

†**Ex. 414.** Prove that the cross-ratio of the range in which four fixed planes through a line cut a variable line is constant.

†**Ex. 415.** The orthogonal projection of the edges of a cube on a plane perpendicular to a diagonal consists of a regular hexagon with three of its diagonals.

†**Ex. 416.** The sum of the squares of the edges of a parallelepiped is equal to the sum of the square of the diagonals.

Ex. 417. Find the angle between two diagonals of a cube.

†**Ex. 418.** Shew that the three diagonals of a regular octahedron are mutually at right angles.

†**Ex. 419.** Prove the following construction for the projection, N, of a point P on a plane α. Take any three points A, B, C on α; draw circles on this plane with centres A, B, C and radii AP, BP, CP. Then N is the radical centre of these circles.

Ex. 420. In a plane making an angle ε with the horizontal is drawn an angle φ one of whose arms is a line of greatest slope; this angle projects into an angle φ′ in the horizontal plane. Prove that tan φ′ = tan φ sec ε.

†**Ex. 421.** Find the centroid of two points whose plans and elevations are given.

†**Ex. 422.** Repeat Ex. 421 for the three vertices of a triangle; for the four vertices of a tetrahedron.

Ex. 423. How must a square prism be cut to give a rhombus of angle 60°?

†**Ex. 424.** The sides AB, BC, CD, DA of a skew quadrilateral are cut by a plane in the points P, Q, R, S respectively. Prove that

$$\frac{AP}{BP} \cdot \frac{BQ}{CQ} \cdot \frac{CR}{DR} \cdot \frac{DS}{AS} = +1,$$

sense being taken into account.

Ex. 425. A solid is made up of a hemisphere and a right circular cone with their bases coinciding. Diameter of hemisphere = diameter of base of cone = 8″. Length of axis of cone = 8¼″.

Draw the plan of the solid when a generator of the cone is in contact with the horizontal plane, and on this plane shew the shape of the section made by a horizontal plane which passes through the centre of the common base of the cone and hemisphere.

†Ex. 426. In a tetrahedron ABCD the straight lines AE, BF, CG, DH pass through a common point P, and meet the faces of the tetrahedron in E, F, G, H respectively; shew that

$$\frac{AP}{AE} + \frac{BP}{BF} + \frac{CP}{CG} + \frac{DP}{DH} = 8.$$

What is the corresponding property of the triangle?

Ex. 427. A man stands on a low roof, with his eye 17 ft. above the ground, opposite a rectangular wall ABDC (fig. 79), of which AB and CD are the vertical ends. He estimates that the angles of elevation of the points A and C are 45° and 20° respectively, that the plane containing his eye and AB makes an angle 45° with a vertical plane perpendicular to the wall, and that the wall is 70 ft. distant from him. Take as axes of x, y, z respectively, the lines through his eye which are (1) perpendicular to the wall, (2) horizontal and parallel to it, and (3) vertical; and find the coordinates of the points A, B, C, D. Find also the area of ABDC in square feet.

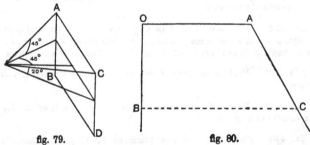

fig. 79.   fig. 80.

Ex. 428. Fig. 80 shews the plan of three walls, BO, OA, AC, having the angle at O a right angle, and the angle at A 120°. OA is 12 feet long, OB is 9 feet long, and BC is perpendicular to OB. The space OACB is to be covered by a plane roof resting on the walls, at a height of 10 ft. from the ground at A and B, and of 8 ft. at O. Find the height of the roof at C.

Taking OA as x-axis, OB as y-axis, and a vertical at O as z-axis, write down the coordinates of the four corners of the roof.

Ex. 429. A ring, radius r, is suspended by a number of equal strings of length $2l$ from an equal ring, fixed vertically above it, the strings being generators of a cylinder. The lower ring is now twisted through an angle θ about a vertical axis and held in the new position, the strings remaining taut, the lower ring still vertically below the upper. Find (i) through what height the lower ring has risen, (ii) the inclination of the strings to the vertical, (iii) their distance from the line joining the centres of the rings.

Ex. 430. If a roof, consisting of two equal rectangles, is 20 feet across horizontally, 30 feet long, and the common (upper) edge of the rectangles is at a vertical height of 12 feet above the parallel (lower) edges, what is the actual area of the two rectangles?

Ex. 431. A square, whose diagonal is 10 inches long, is placed inside a hollow sphere of radius 13 inches, the four corners of the square resting on the inner surface of the sphere.

Draw (i) a section containing a diagonal of the square and passing through the centre of the sphere; (ii) a section containing a side of the square, and perpendicular to the plane of the square. [Scale $\frac{1}{4}$ full size.]

Find the perpendicular distance of the plane of the square from the centre of the sphere.

Ex. 432. A church spire is in the form of a cone 30 feet high and 10 feet in diameter at the base. Supposing the sun's altitude to be $55°$, draw a plan shewing the base of the spire and the shadow which it would throw on a horizontal plane through its base. Find what proportion of the slant surface of the spire is in shadow.

Ex. 433. The two faces of a burning glass are parts of spheres of radii a and b, and the diameter of the glass is c. Find an expression for the thickness of the glass at the centre.

SOME PROPERTIES OF THE TETRAHEDRON.

†**Ex. 434.** If opposite edges of a tetrahedron are equal, its faces are congruent.

†**Ex. 435.** If the opposite edges of a tetrahedron be equal two and two, prove that the faces are acute-angled triangles. Prove also that a tetrahedron can be formed of any four congruent acute-angled triangles.

†**Ex. 436.** A sphere can be circumscribed to a tetrahedron. The line joining its centre to the mid-point of an edge is perpendicular to that edge; the line joining its centre to the circumcentre of a face is perpendicular to that face.

†Ex. 487. Eight spheres (in general) exist, each of which touches the four planes forming the faces of a tetrahedron.

†Ex. 488. Eight spheres (in general) exist, each of which touches the four sides of a skew quadrilateral (produced if necessary). If one of these spheres touches all four sides within their unproduced portion, the sums of opposite sides are equal.

†Ex. 489. If a sphere exists which touches all the edges of a tetrahedron within their unproduced portion, the three sums of opposite edges are equal. What is the intersection of this sphere with a face?

†Ex. 440. Prove that G, the centroid of the four vertices of a tetrahedron, lies ¾ of the way from each vertex to the centroid of the opposite face. Shew that the lines joining mid-points of opposite edges cointersect and are bisected at G.

†Ex. 441. Prove that any plane parallel to two opposite edges of a tetrahedron cuts the faces in a parallelogram, whose angles are constant for different planes. Also that if the two edges are at right angles, the section may be a square.

Ex. 442. In the case of a regular tetrahedron, compare the radii of the circumscribed sphere, the inscribed sphere, and the sphere touching the six edges at their mid-points.

†Ex. 448. If from the orthocentre of a face of a tetrahedron a line be drawn perpendicular to that face it will intersect the perpendiculars to the other three faces drawn from the opposite corners.

TETRAHEDRON AND ASSOCIATED PARALLELEPIPED.

Through each pair of opposite edges of a tetrahedron construct a pair of parallel planes; this gives a parallelepiped circumscribed to the tetrahedron, the edges of the latter being diagonals of the faces of the former.

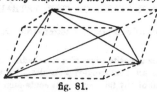

fig. 81.

Ex. 444. What is the parallelepiped associated with a regular tetrahedron?

†Ex. 445. Use the associated parallelepiped to prove that joins of mid-points of opposite edges cointersect and bisect one another.

†Ex. 446. ABCD is a tetrahedron, AA′ is a diagonal of the associated parallelepiped. Prove that AA′ passes through the centroid of BCD. In what ratio is AA′ cut?

†Ex. 447. If four points be so situated that the distance between each pair is equal to the distance between the other pair, prove that the angles subtended at any one of these points by each pair of the others are together equal to two right angles.

†Ex. 448. If each edge of a tetrahedron be equal to the opposite edge, the straight line, joining the middle points of any two opposite edges, is at right angles to each of those edges.

Ex. 449. Prove that the volume of the parallelepiped is three times that of the tetrahedron; and hence that the volume of a tetrahedron is $\frac{1}{6}$ × product of opposite edges × distance between them × sin (their inclination).

†Ex. 450. If six planes be constructed, each of which contains one edge and bisects the opposite edge of a tetrahedron, these planes have a common point.

†Ex. 451. Shew that, in any tetrahedron, the line joining the mid-points of one pair of opposite edges is perpendicular to the shortest line between either of the other pairs of opposite edges.

†Ex. 452. If two of the joins of mid-points of opposite edges of a tetrahedron are at right angles, the remaining edges are equal.

†Ex. 453. Shew that if the sum of the squares of two opposite edges of a tetrahedron is equal to the sum of the squares on another pair of opposite edges, the two remaining opposite edges are at right angles to one another.

†Ex. 454. Shew that the shortest distance between two opposite edges of a regular tetrahedron is equal to half the diagonal of the square described on an edge.

†Ex. 455. In a tetrahedron, of which the opposite edges are equal, shew that (i) the centroid, the centre of the inscribed sphere, and the centre of the circumscribing sphere all coincide, (ii) the shortest distance between opposite edges bisects these edges, (iii) the three shortest distances are all mutually at right angles and are bisected in the centroid, (iv) the sum of the squares on the sides of a face is double the square on the diameter of the circumscribing sphere.

ORTHOCENTRIC TETRAHEDRON.

†**Ex. 456.** Prove that the perpendiculars from A upon BCD and from D upon ABC do not intersect unless the edges AD and BC are at right angles.

†**Ex. 457.** If two pairs of opposite edges of a tetrahedron are at right angles, the third pair also are at right angles.

†**Ex. 458.** Prove that, if the opposite edges of a tetrahedron are at right angles, the four altitudes will meet in a point.

Such a tetrahedron has an orthocentre, and is called an **orthocentric tetrahedron.**

Ex. 459. What is the parallelepiped associated with an orthocentric tetrahedron?

†**Ex. 460.** Prove that a vertex of an orthocentric tetrahedron projected orthogonally on to the opposite face gives the orthocentre of that face.

†**Ex. 461.** Shew that the three common perpendiculars of opposite edges of an orthocentric tetrahedron meet at the orthocentre.

†**Ex. 462.** Prove that if a, a; b, β; c, γ be pairs of opposite edges of an orthocentric tetrahedron, then $a^2 + a^2 = b^2 + \beta^2 = c^2 + \gamma^2$.

†**Ex. 463.** In a tetrahedron each edge is perpendicular to the direction of the opposite edge; prove that the straight line joining the centre of the sphere, circumscribing the tetrahedron, to the middle point of any edge, is equal and parallel to the straight line joining the centre of perpendiculars to the middle point of the opposite edge.

†**Ex. 464.** Prove (i) that the joins of mid-points of opposite edges of an orthocentric tetrahedron are equal; (ii) hence that the mid-points of the six edges lie on a sphere whose centre is the centroid; (iii) that the sphere passes through the feet of the perpendiculars from vertices upon edges; (iv) that its centre is midway between the orthocentre and the circumcentre of the tetrahedron, and projects orthogonally into the nine-point centre of each face.

†**Ex. 465.** If the line joining a vertex A of an orthocentric tetrahedron to the orthocentre O cuts the opposite face in S and the circumscribing sphere in Σ, then $S\Sigma = 2OS$.

†**Ex. 466.** If E is the orthocentre of an orthocentric tetrahedron ABCD, shew that any one of the five points A, B, C, D, E is the orthocentre of the tetrahedron determined by the other four.

INVERSION.

†Ex. 467. A plane through the centre of inversion inverts into a plane.

†Ex. 468. A plane not through the centre of inversion inverts into a sphere through the centre of inversion; the perpendicular from the centre of inversion upon the plane passes through the centre of the sphere.

†Ex. 469. A sphere through the centre of inversion inverts into a plane.

†Ex. 470. A sphere not through the centre of inversion inverts into a sphere.

†Ex. 471. If the tangents from the centre of inversion to two given spheres are equal, the spheres can be inverted into themselves. The locus of such centres is a plane.

†Ex. 472. A circle is inverted with respect to a sphere whose centre O does not lie in the plane of the circle; prove that the inverse is a circle, and shew that the point P which inverts into the centre of the inverse circle is obtained thus: Describe a sphere through O and the circumference of the given circle; join O to the pole of the plane of the circle with respect to this sphere; this line cuts the sphere at P.

†Ex. 473. Two intersecting curves in space cut at the same angle as their inverses.

†Ex. 474. A sphere is inverted from a point on its surface; shew that to a system of parallels and meridians on the surface will correspond two systems of coaxal circles in the inverse figure.

†Ex. 475. Prove that, if P, Q be the ends of a diameter of a small circle of a sphere, O a point of the great circle PQ, and R any point on the small circle, then the arcs of the small circles PRO, RQO are perpendicular to each other at R.

†Ex. 476. Circles are drawn to cut a given circle orthogonally at two points of intersection, and to pass through a given point not in the plane of the circle. Shew that they intersect in another common point; and hence shew how a circle and a point not in its plane may be inverted respectively into circle and centre.

†Ex. 477. Shew that the locus of points with respect to which an anchor ring can be inverted into another anchor ring consists of a straight line and a circle.

POLE AND POLAR.

†Ex. **478.** Prove that the tangent lines from a point to a sphere form a right circular cone, which touches the sphere along a small circle.

*Such a cone is called an **enveloping cone** of the sphere.*

†Ex. **479.** If the plane of the small circle in which the enveloping cone from a point T touches a sphere (centre C) cut CT in N ; then CT is perpendicular to the plane, and CT . CN = (radius of sphere)².

*If a point T and a plane are so related that the perpendicular CN from the centre C of a sphere to the plane passes through the point, and CN . CT = (radius)², then the point and plane are called **pole** and **polar plane** with respect to the sphere.*

†Ex. **480.** If a straight line pass through a point, the tangent planes at the points where the line cuts a sphere intersect on the polar plane of the point.

†Ex. **481.** If a plane be drawn through a point to cut a sphere in a circle, the enveloping cone which touches the sphere along the circle has its vertex on the polar plane of the point.

†Ex. **482.** If the polar plane of P passes through Q, the polar plane of Q passes through P.

†Ex. **483.** If a variable point lies on a fixed line, the polar plane of the point contains another fixed line.

Shew that the relation between these lines is reversible.

*Two lines so related are said to be **polar lines.***

†Ex. **484.** Polar lines are at right angles, their common perpendicular passes through the centre, and the product of their distances from the centre is equal to the square of the radius.

†Ex. **485.** What is the polar line of a line touching a sphere?

†Ex. **486.** A plane through any point cuts a sphere and the polar plane of the point in a circle and the polar line of the point with respect to the circle.

†Ex. **487.** An orthocentric tetrahedron has a real or imaginary **polar sphere** whose centre is at the orthocentre; i.e. a sphere with respect to which each vertex is the pole of the opposite face, and each edge is the polar line of the opposite edge. This sphere cuts each face in the real or imaginary polar circle of the triangle.

†Ex. **488.** It is assumed that the student is familiar with the properties of the **radical axis** of two circles, and the **radical centre** of three circles. Investigate analogous properties for spheres.

†Ex. **489.** Investigate the properties of the **centres of similitude** of two spheres.

ANSWERS TO NUMERICAL EXAMPLES.

Answers are usually given to 4 significant figures; as 4-figure tables have been used, the 4th figure is liable to error.

72. 35° 16′.   **77.** (i) 70° 32′, (ii) 109° 29′.   **79.** 59° 12′.

80. 13° 48′.   **81.** 18° 45′, 33° 49′.   **82.** 95° 52′.

83. 51° 59′, 42° 7′.   **85.** (i) 14° 29′, (ii) 7° 58′, (iii) 12·77″, (iv) 12·41″.

86. 45°, 35° 16′, 1″.   **87.** AF = 3″, BF = 2·236″, tan = $\frac{1}{2}$.

96. 73° 44′.   **132.** (i) 84 sq. cm., (ii) 168 sq. cm.   **134.** 31·31 sq. in.

138. (a) (i) $\sqrt{\frac{1}{3}S}$, (ii) $\sqrt{\frac{1}{2}S}$; (b) (i) $V^{\frac{1}{3}}$, (ii) $3^{\frac{1}{4}}V^{\frac{1}{3}}$.   **139.** $6V^{\frac{2}{3}}$.

141. 8 cu. in.   **142.** 93677 cu. m.   **143.** 4·6 cu. ft.

144. 206·5 sq. ft., 288 cu. ft.   **145.** After 10 hours.

156. $\sqrt{\pi}$: 2 = ·886 : 1.   **157.** $\dfrac{4840 \times 9}{2\pi r h}$.   **160.** $2\dfrac{V}{h} + 2\sqrt{\pi V h}$.

161. 4 : π.   **162.** 21·5 %.   **164.** 66·8 lbs.   **166.** $\pi = 2·9$.

167. $x = 3·42$, weight = 9·145 lbs.   **173.** 4·796″, (i) 78° 13′, (ii) 73° 34′.

174. $\sqrt{b^2 - a^2}$, (i) $\cos^{-1}\dfrac{a\sqrt{3}}{\sqrt{4b^2 - a^2}}$, (ii) $\cos^{-1}\dfrac{a}{b}$.

175. $\sqrt{\frac{2}{3}}$: 1 = ·8165 : 1, 70° 32′.

177. (i) 80° 10′, (ii) 62° 52′, (iii) 70° 54′, (iv) 28° 56′.

178. (i) 8·718 ft.,   (ii) 38° 15′.   **180.** $S + S\dfrac{\sqrt{h^2 + r^2}}{r}$.

182. $\frac{7}{12}x^3$ cu. ft.   **183.** 174 tons.   **189.** 54° 59′.   **190.** $\cos^{-1}\dfrac{d}{2l}$.

195. $\pi r^2 + \pi r\sqrt{r^2 + h^2}$.   **196.** $\dfrac{n^2 - 1}{n^2}S$.   **197.** $\frac{1}{3}\pi r^2 h$.

198. (i) $\frac{1}{3}\pi h^3 \tan^2 a$, (ii) $\frac{1}{3}\pi (s^2 - h^2) h$.

200. Surfaces 36π, 24π, 16·8π sq. in.; Volumes 16π, 12π, 9·6π cu. in.

202. 566·3 sq. ft.; 1571 cu. ft.   **203.** 36·40 cu. in.   **204.** $\pi : 3\sqrt{3}$.

206. 45·93 sq. in.   **207.** Surface $= \pi (a^2 + b^2) + \pi (a + b) \sqrt{(a - b)^2 + h^2}$;

208. When $x = \dfrac{1}{\sqrt{3}}$, $2\pi \sqrt{\frac{2}{3}}$.   Volume $= \frac{1}{3}\pi h (a^2 + ab + b^2)$.

216. $\pi(r^3 - h^3)$.   **218.** $2\sqrt{r^3 - d^3}$.   **221.** $\pi\,\dfrac{r(l^3 - r^3)}{l}$.

224. $73\frac{1}{2}$ miles.   **226.** $6 \cdot 582$ cm.   **233.** $r\theta$, $r\dfrac{x}{180}\pi$.

238. $\cos^{-1}\dfrac{a}{r}$, $\dfrac{2a\sqrt{r^3 - a^3}}{r}$.   **240.** $\cdot 6497 : 1$.   **241.** $1 \cdot 016 : 1$, $20 \cdot 94$ miles.

242. $a : 2\pi$.   **243.** Each angle and side $= 90°$, area $= \frac{1}{8}$ area of sphere.

244. $13 \cdot 27$ in., $a = c = 35° 6'$, $b = 50° 28'$, $B = 95° 44'$.   **246.** (i) $\pi : 6$, (ii) $\pi : 6$.

247. $1 \cdot 241 : 1$.   **248.** $1 \cdot 588 : 1$.   **250.** $\cdot 241$ cu. in.   **253.** $\cdot 6181''$.

255. $\cdot 972$.   **257.** $\dfrac{l - r}{2l}$.   **258.** Volume $= \frac{7}{12}\pi r^3$, area $= \frac{5}{2}\pi r^2$.

260. $2 \cdot 5$ m., $7 \cdot 069$ sq. m.   **261.** $2 \cdot 82$ in.   **264.** $\dfrac{a^2}{6h} + \dfrac{h}{2}$.

265. $2\sqrt{ab}$, where a and b are the radii of the ends.

271. (i) 3, (ii) 1, (iii) 1, (iv) 0.   **276.** (i) 3, (ii) 1, (iii) 1, (iv) 0.

279. 3 or 4.   **280.** Edge $\times \sqrt{2}$.   **292.** 3.   **296.** 6.   **297.** $\sqrt{38}$; $\sqrt{53}$.

298. (i) $\sqrt{29}$, (ii) $\sqrt{(a - p)^2 + (b - q)^2 + (c - r)^2}$.

299. $\sin^{-1}\frac{2}{3} = 41° 49'$, $\sin^{-1}\frac{1}{3} = 19° 28'$, $\sin^{-1}\frac{2}{3} = 41° 49'$; $\cos^{-1}\frac{2}{3} = 48° 11'$, $\cos^{-1}\frac{1}{3} = 70° 32'$, $\cos^{-1}\frac{2}{3} = 48° 11'$.

315. $4 \cdot 717$ ft.; $(3\frac{1}{2}, 7\frac{1}{4}, 0)$; $(5\frac{4}{9}, 0, 6\frac{3}{14})$; $39° 30'$.   **323.** $4 \cdot 653$ cm.

328. $17 \cdot 12$ ft.   **329.** $3 \cdot 421''$.   **330.** $17 \cdot 12$ ft.   **331.** $1 \cdot 732$ m.

332. $3 \cdot 833$ in.   **333.** $2 \cdot 8$ in.   **340.** $\dfrac{c}{f} = \dfrac{b}{d} = \dfrac{a}{a - e}$.

343. $50° 12'$, $44° 30'$.   **345.** (ii) $32° 48'$, (iii) $62° 58'$.

346. Horizontal projection $= 2 \cdot 458''$, vertical $= 1 \cdot 881''$.

347. Horizontal projection $= 6 \cdot 578$ cm., vertical $= 3 \cdot 812$ cm.

348. $36° 52'$, $48° 56'$.   **356.** $2 \cdot 32''$, $3 \cdot 96''$, $1 \cdot 99''$.   **358.** $47° 16'$, $25° 5'$.

359. $40° 7'$, $57° 15'$, $18 \cdot 49$ ft.   **360.** $5 \cdot 79$ ft., $26° 2'$.   **361.** (i) $15 \cdot 57$ ft., $24° 40'$; (ii) $50° 2'$; (iii) $134° 42'$.   **363.** (i) $3 \cdot 075$ in., (ii) $40° 5'$.

364. N. $53° 8'$ E.; $39° 48'$.   **365.** $4 \cdot 717$ ft.; $(3\frac{1}{2}, 7\frac{1}{4}, 0)$; $(5\frac{4}{9}, 0, 6\frac{3}{14})$; $39° 30'$.

368. $123 \cdot 6$ cm.   **369.** $65° 56'$.   **370.** $72° 21'$.   **372.** $4 \cdot 082$ cm.

373. $7 \cdot 5$ cm.   **374.** $1 \cdot 06$ in.   **417.** $70° 32'$.

427. A, $(70, -70, 99)$; B, $(70, -70, -17)$; C, $(70, 263, 99)$; D, $(70, 263, -17)$; area $= 38,600$ sq. ft.

428. $12 \cdot 9$ ft. $(0, 0, 8)$ $(12, 0, 10)$ $(17 \cdot 2, 9, 12 \cdot 9)$ $(0, 9, 10)$.

429. (i) $2l - 2\sqrt{l^2 - r^2 \sin^2\dfrac{\theta}{2}}$, (ii) $\sin^{-1}\left(\dfrac{r}{l}\sin\dfrac{\theta}{2}\right)$, (iii) $r\cos\dfrac{\theta}{2}$.

430. 937 sq. ft.   **431.** 12 in.   **432.** $\cdot 423$ of the whole surface.

433. $a + b - \sqrt{a^2 - \dfrac{c^2}{4}} - \sqrt{b^2 - \dfrac{c^2}{4}}$.

442. $3 : 1 : \sqrt{3}$.   **446.** $2 : 1$.

INDEX.

Printed in the United States
By Bookmasters